Risk Management in Supply Chains

The book examines a relatively unexplored issue in supply chain risk management, which is how long companies specifically take to respond to catastrophic events of low probability but high impact. The book also looks at why such supply chain disruptions are unavoidable, and consequently, all complex supply chains are inherently at risk.

The book illustrates how companies can respond to supply chain disruptions with faster responses and in shorter lead-times to reduce impact. In reducing total response time, designing solutions, and deploying a recovery plan sooner after a disruption in anticipation of such events, companies reduce the impact of disruption risk. The book also explores the basics of multiple-criteria decision-making (MCDM) and analytic hierarchy process (AHP), and how they contribute to both the quality of the financial economic decision-making process and the quality of the resulting decisions. The book illustrates through cases in the construction sector how this industry has become more complex and riskier due to the diverse nature of activities among global companies.

Mohammad Heydari is a PhD student in Management Science and Engineering at the School of Economics and Management, Nanjing University of Science and Technology, China.

Kin Keung Lai is Professor at the College of Economics, Shenzhen University, China.

Zhou Xiaohu is Professor at the School of Economics and Management, Nanjing University of Science and Technology, China.

Routledge Advances in Risk Management

Edited by Kin Keung Lai and Shouyang Wang

Operational Risk Management in Container Terminals
Eric Su, Edward Tang, Kin Keung Lai, Yan Pui Lee

Managing Currency Options in Financial Institutions
Vanna-Volga Method
Yat-Fai Lam and Kin-Keung Lai

Gold and International Finance
The Gold Market Under the Internationalization of RMB in Hong Kong
Haywood Cheung

Green Transportation and Energy Consumption in China
Jian Chai, Ying Yang, Quanying Lu, Limin Xing, Ting Liang, Kin Keung Lai and Shouyang Wang

Chinese Currency Exchange Rates Analysis
Risk Management, Forecasting and Hedging Strategies
Jiangze Du, Jying-Nan Wang, Kin Keung Lai, and Chao Wang

Forecasting Air Travel Demand
Looking at China
Yafei Zheng, Kin Keung Lai and Shouyang Wang

Supply Chain Risk Management in the Apparel Industry
Peter Cheng, Yelin Fu and Kin Keung Lai

Risk Management in Supply Chains
Using Linear and Non-linear Models
Mohammad Heydari, Kin Keung Lai and Zhou Xiaohu

For more information about this series, please visit www.routledge.com/Routledge-Advances-in-Risk-Management/book-series/RM001

Risk Management in Supply Chains

Using Linear and Non-linear Models

Mohammad Heydari, Kin Keung Lai
and Zhou Xiaohu

Routledge
Taylor & Francis Group
LONDON AND NEW YORK

First published 2020
by Routledge
2 Park Square, Milton Park, Abingdon, Oxon OX14 4RN

and by Routledge
52 Vanderbilt Avenue, New York, NY 10017

Routledge is an imprint of the Taylor & Francis Group, an informa business

British Library Cataloguing-in-Publication Data
A catalogue record for this book is available from the British Library

Library of Congress Cataloging-in-Publication Data
A catalog record has been requested for this book

ISBN: 978-0-367-35951-5 (hbk)
ISBN: 978-0-429-34282-0 (ebk)

Typeset in Galliard
by codeMantra

Contents

Figures

Tables

Preface

A wide range of recent man-made and natural disasters has demonstrated the importance of managing disruption risk in global supply chains. This research argues that supply chain disruptions are, de facto, unavoidable, and consequently, all complex supply chains can be considered inherently risky. This research focuses on a relatively unexplored issue in supply chain risk management, asking and answering the question of how companies specifically use time to respond to catastrophic events of low probability but high impact. Linking faster response lead-time with reduced impact, the goal is to identify and explore the underlying factors of managing disruption risk by answering how companies respond to supply chain disruptions. In reducing total response time by detecting the event, designing solutions, and deploying a recovery plan sooner after a disruption, the company can reduce the impact of disruption risk.

Efforts to describe and explain supply chain management (SCM) have recently led to a plethora of research and writing in this field. At the same time, the level of attention SCM now receives in business practices also heavily influences the growing interest in SCM research. SCM is now seen as a governing element in strategy (Fuller, O'Conner and Rawlinson, 1993) and an effective way of creating value for customers. However, despite the growing interest in SCM, there is still a lack of cohesive information that explains the SCM concept and emphasizes the variety of research work being accomplished in this area.

Other researchers have also provided taxonomies and frameworks that help both practitioners and academics understand how to manage supply chains. For example, Bowersox (1969), in addition to reviewing relevant streams of thought in physical distribution, suggested that the distribution function can provide a competitive advantage through channel-wide integration beyond the firm. Shapiro (1984) provided a prescriptive framework that can enable a company to gain leverage by ensuring a good fit between its logistics system and competitive strategy. Houlihan (1985) made a strong case for viewing the supply chain as a single entity by incorporating a logistics focus into the strategic decisions of the firm. Langley (1992) cast the evolution of logistics into three specific contexts: past (1950–1964), present (1965–present), and future. Since that time, a variety of authors [Stevens, 1989, Masters and Pohlen, 1994, Mourits and Evers, 1995, and Thomas and Griffin, 1996] have added to this body of literature by providing integrative frameworks to help design and manage supply chains.

Despite the efforts of previous authors, we believe that the growing literature in SCM warrants a close re-examination of published SCM works to date. The purpose of such an update is to carefully chart the historical development of SCM and synthesize future directions of research. Simply starting off with a literature search into this SCM field quickly becomes overwhelming due to the amount of work being done that seemingly falls into the subject. Therefore, it would be beneficial, to both the novice and even those familiar with the subject, for us to discuss, describe, and categorize the work being done in supply chain management.

In Chapter 1, we explore the basics of MCDM and AHP. This special issue is devoted to MCDM and finance. For people working in finance, in academia, in practice, or in both, the combination of 'finance' and 'multiple criteria' is not obvious. For many years, we have been involved in the study and practice of financial economic decision-making. Our main focus was and is on financial management science: the development of tools and frameworks to support financial decision-making. We are convinced that many financial decision problems are in fact multiple-criteria decision problems. In addition, we believe that many of the tools developed in the field of MCDM contribute both to the quality of the financial economic decision-making process and to the quality of the resulting decisions. In the present book, we answer the question of why financial decision problems should be considered as multiple-criteria decision problems and treated accordingly.

In Chapter 2, we explore recent decades; projects in the construction sector have become more complex and riskier due to the diverse nature of activities among global companies (Akintoye, A.S.; MacLeod, M.J., 1997; Ribeiro, C.; Ribeiro, A.R.; Maia, A.S.; Tiritan, M.E., 2017). In comparison to other sectors, construction projects encounter more risks due to uncertainties occurring because of various construction practices, working conditions, mixed cultures, and political conditions between host and home countries (Iqbal, S.; Choudhry, R.; Holschemacher, K.; Ali, A.; Tamošaitien˙ e, J., 2015; Pak, D.; Han, C.; Hong, W.-T., 2017). Thus, in this scenario, risk management can be considered a vital part of the decision-making process in construction projects. These projects may involve many stakeholders, in addition to uncertain socio-economic conditions at the project site, bringing big challenges to practitioners of the industry in recent decades (Ravanshadnia, M.; Rajaie, H., 2013; Ebrat, M.; Ghodsi, R., 2014). Construction project failure may cause higher costs and time over-runs, requiring a systematic risk assessment and evaluation procedure to classify and respond to changes (Taylan, O.; Bafail, A., 2014; Dziadosz, A.; Rejment, M., 2015).

Thus, prioritization among construction-based risk portfolios, and finding suitable risk mitigation strategies for construction projects, can be introduced as MCDM problems. Researchers have recently proposed new methods for prioritizing risks in construction-based projects (Schieg, M., 2006; Santos, R.; Jungles, A., 2016; Hwang, B.-G.; Zhao, X.; Yu, G.S., 2016). In addition, the increasing dynamism of construction projects has resulted in extensive imprecise-ness and subjectivities in this risk investigation procedure. With respect to the identification of risk criteria, a methodology is needed to sort and priorities criteria weights, based on specific environments and domain experts' judgment. In the real world, since various uncertainties occur in the decision-making process due to the subjective and qualitative judgment of decision makers (DMs), it is essential to develop a more optimized technique that can handle various types of uncertainties (Sadeghi, N.; Fayek, A.; Pedrycz, W., 2010; Taylan, O.; Bafail, A.; Abdulaal, R.; Kabli, M., 2014; Vafadarnikjoo, A.; Mobin, M.; Firouzabadi, S., 2016).

Chapter 3 demonstrates how the usual methods of modeling risk and uncertainty, which are inherent in agricultural decision-making, can be incorporated within the MCDM framework. It is also shown that the traditional risk and uncertainty analysis, by its very nature, is multi-objective analysis, involving two objectives: profit and a measure of its variability. Treating the risk and uncertainty models as particular cases of the MCDM paradigm has therefore both theoretical and practical advantages. It is possible to create a hybrid of the Markowitz and mean absolute deviations approaches with compromise programming to obtain compromise sets, which lie closest to an ideal point defined in terms of an acceptable level of risk. Similarly, the game-theoretic approach can be extended to what may be called compromise games, enabling us to explain the behavior of the DM by considering a set of conflicting criteria rather than relying on the naïve assumption of a single criterion optimization by DMs.

In Chapter 4, we explore the basics of SCM from a conceptual perspective by tracing the roots of the definition and the origins of the concept from a broad stream of literature. Recognizing that there is not a clear consensus on the definition of supply chain management, we look at the various approaches to defining it, from the 1980s to the present, and provide our own interpretation based on the literature. Our working definition of a supply chain is from Stevens (1989), who defines it as:

"...a connected series of activities which is concerned with planning, coordinating and controlling materials, parts, and finished goods from supplier to customer. It is concerned with two distinct flows (material and information) through the organization."

In Chapter 5, we show that the paths leading to the current state of SCM have evolved over the past four decades. In particular, we characterize SCM as evolving over the years from models in supply chain management. We also show that it benefits from a variety of concepts that were developed in several different disciplines, including marketing, economics, operations research, management science, operations management, and logistics.

In Chapter 6, we summarize the volume of SCM research into three broad categories – competitive strategy, firm-focused tactics, and Environmental Management and operational efficiencies – based on the level and detail of the SCM problems being addressed.

In Chapter 7, we focus on the research methodologies and solution approaches that have been used to address SCM problems. In particular, we categorize these research methodologies into four broad categories: concepts, case-oriented, frameworks, and models.

Finally, in the Appendix, we provide a database of selected papers to help summarize the volume of research that has been done in the SCM arena.

The research identifies four categories of factors that companies can focus on to reduce response time in the face of catastrophic events of low probability and high impact: *organizational structure, preparation, partnership, and reserve.* The research derives new insights, presented as four propositions that relate the response time in managing supply chain disruption to negative or potentially positive impact.

We have intentionally limited our focus to articles that have already been published at the time of this writing as an attempt to understand the state of research at a fixed point in time. Naturally, in an emerging field like SCM, there is much research in the pipeline, and a new "chapter" is needed periodically. We hope this is useful for today's researchers.

Mohammad Heydari
Kin Keung Lai
Zhou Xiaohu

1 Introduction

1.1 Introduction

The necessity of quantitative estimation of non-failure operation of complex technical structures at the beginning of the 60s XX centuries stimulated the so-called logic and probabilistic calculus (LPC) which is a part of the mathematics treating rules of calculus and operating with statements of two-valued logic. LPC is based on the logic algebra and rules of replacement of logical arguments in functions of the logic algebra (PAL) by probabilities of their being true and rules of replacement of the logic operations by the arithmetic ones.

In other words, with the help of LPC, it became possible to connect the Boolean algebra with the probability theory not only for the elementary structures but also for the structures, whose formalization results in PAL of iterated type (bridge, network, monotonous). This original "bridge of knowledge" includes some proven theorems, properties, and algorithms, which constitute the mathematical basis of LPC.

Investigation of the safety problem has resulted in the development of the original logic and probabilistic theory of safety (LPTS), which allows to estimate quantitatively the risk of system (as a measure of its danger) and to rank the contribution of separate arguments to the system danger (in the case of an absence of truth probabilities of initiating events). The ranking of arguments under their contribution to the system reliability was proposed by me in 1976 in the monograph [Reliability of Engineering Systems. Principles and Analysis. Mir Publicizes, Moscow, 1976, p. 532] with the help of the introduction of concepts: "Boolean difference,""weight," and "importance" of an argument. The aim of the author, from my point of view, is the connection of the LPC used in the field of technical systems, with questions of risk in economics and organizational systems.

Studying the works by the author, I realized that these economic and organizational systems essentially differ from technical ones, and the direct carrying of the knowledge and results of LPC from the area of engineering into an area of economics is not effective, and sometimes, it is not even possible. It is likely that much time and many efforts will be needed so that the new approaches in the LPC can make the same revolutionary break in the financial market, which was made by George Bool in the development of the inductive logic in the middle of XIX century and by G. Markowitz in the choice of the optimal security portfolio with the help of the analytical theory of probabilities in the middle of XX century.

To the author's knowledge, the risk phenomenon in complex technical, economic, and organizational systems is not completely recognized in the scientific plane and is also not resolved satisfactorily for the needs of applications, even though in complex systems, non-success occurs rather often with human victims and large economic losses. The management risk problem is current and challenging; it forces us to carry out new investigations and to seek new solutions for quantitative estimation and analysis of risk.

The risk is quantitative measure of fundamental properties of systems and objects, such as safety, reliability, effectiveness, quality, and accuracy. The risk is also a quantitative measure of non-success of such processes and actions as classification, investment, designing, tests, operation, training, development, management, etc.

In the listed subject fields, we shall consider three different statements of mathematical tasks of optimization by the management of risk – of interest will be the risk in problems of classification, investment, and effectiveness. Generally, the riskis characterized by the following quantitative parameters:

- The probability of non-success.
- The admitted probability of non-success (admitted risk).
- Maximum admitted losses or minimal admitted effectiveness.
- Value of losses or the effectiveness parameter.
- The number of different objects or conditions of the object in the system.
- The number of dangerous objects or conditions of an object.

It was marked by the founders of many fields of modern science, John von Neumann and Norbert Wiener, that the behavior of complex technical, economic, and social systems cannot be described with the help of differential equations. However, the description can be madeby the logic and the set theory, instead of the theories of chaos, accidents, bifurcations, etc. (See the book by Morgenstern and Neumann "The game theory and economic behavior," Moscow, Nauka, 1970, sec. 1.2.5. and 4.8.3.)

Analysis of the theories of Management and Risk development and the interaction between Man and Risk in complex systems proves the correctness of this point of view. In complex human-machine systems, the logic and probabilistic theory (LP-theory) reveals considerable achievements in estimation, analysis, and forecasting of risk (Ryabinin,1976; Guding et al., 2001).

The LP-theory attractiveness is in its exclusive clearness and unambiguity in quantitative estimations of risk; in a uniform approach to risk problems in economics and engineering; and in big opportunities for the analysis of influence by any element, including personnel, on the reliability and safety of the whole system. The risk LP-model may include the logic connections OR, AND, NOT between elements of system and cycles. Elements of the system under consideration may have several levels of conditions. The system risk dynamics can be taken into account by consideration of variation in time of probabilities of conditions.

The basis for the construction of the scenario risk LP-management in complex systems is the risk LP-theory; the methodology for the construction of scenarios and models of risk; the technology of risk management; examples of risk modeling and analysis from various fields of economics and engineering.

In complex systems, the technology of the scenario risk LP-management is based on the risk estimation by the LP-model, the techniques of the risk analysis, schemes and algorithms of risk management, and the corresponding software. Generally, it is impossible to control the risk without a quantitative analysis of risk which allows us to trace the contributions of initial events to the risk of the system. Estimation and analysis of risk as well as finding optimal management are carried out algorithmically with calculations, which are very time-consuming even for modern computers.

The risk LP-theory considered in the book unifies Ryabinin's LP-calculus and LP-method, Mojave's methodology of automatized structure and logical modeling, and Solojentsev's risk LP-theory with groups of incompatible events (GIE[1]). The LP-calculus is a special part of discrete mathematics, which should not be confused with the probabilistic logic and other sections of the mathematical logic. Therefore, it is useful to outline the history of the publications on this subject briefly. To the author's knowledge, the idea and development of the subject should be attributed to Russian authors. The contents and formation of LP-calculus originate from the work by I. A. Ryabinin "Leningrad scientific school of the logic and probabilistic methods of investigations of reliability and safety" (in the book: "Science of St. Petersburg and sea power of Russia," v. 2, 2002, pp. 798–812).

The LP-calculus was createdat the beginning of the 60-the of XX century in connection with the necessity of quantitative estimation of the reliability of complex structures (annular, networks, bridge-like and monotonous ones). Scientific literature of that time could suggest nothing suitable to deal with the problem. The experts in reliability could perform calculations for the consecutive, parallel, or tree-like structures only.

In 1987, Kyoto University published the book by I. A. Ryabinin and G. N. Cherkesov "Logic and probabilistic methods of research of reliability structural-complex systems" (M.: Radio and Communication, 1981, p. 264) translated into the Japanese language. In the book, the set-theoretic and logic part of LP-calculus was advanced. In the new book "Reliability and safety of structural-complex systems" (SPb., Polytechnika, 2000, p. 248), Prof. I. A. Ryabinin has generalized the 40-year experience of researches on reliability and safety by the LP-calculus. There is a review of this book in English (Andrew Adamatzky "Book reviews"– Reliability and Safety of Structure-complex Systems. – Kybernetes. Vol. 31, No 1, 2002, pp. 143–155).

The present publications in the risk LP-theory and the risk management do not represent the state-of-the-art in the field of science; they have a small circulation, and the knowledge is confined within a small group of experts. The risk LP-theory and such scientific disciplines as the LP-calculus, the discrete mathematics, and the combinatorial theory are not included as a rule into the educational programs of the Higher School. It causes the difficulty in the way of active mastering the scenario risk LP-management in business, economics, and engineering. The publication of the present monograph, devoted to the scenario risk LP-management, seems to be well-timed.

The present book is of applied importance. The purpose of the present book is to acquaint economists, engineers, and managers with the bases of the scenario risk LP-management, which includes: the risk LP-theory, the methodology of construction of the risk scenario, the technology of risk management, examples of scenarios, and models of risk in different fields of economy and engineering.

The important feature of the suggested presentation is the attempt to unify knowledge from different fields: discrete mathematics, combinatorial theory and Weil's theorem; nonlinear optimization and algorithmic calculations, modeling of Monte-Carlo and on modern computers; the LP-calculus (Ryabinin,1976; Guding et al., 2001), the LP-methods (Mojaev and Gromov, 2000; Ryabinin, 2000); the theories by Markowitz and VaR for risk of security portfolio (Markowitz, 1952; Sharp, 2001), the risk LP-theory with GIE (Solojentsev et al., 1999; Solojentsev and Alekseev, 2003).

The novelty and utility of the book consist of the following: it is the first time when the basic principles of the modern risk LP-theory (the LP-calculus, the LP-methods, and the risk LP-theory with GIE) are stated in one work using uniform methodology and terminology, and with practical orientation on use both in engineering and economics. With the permission of Prof. I. A. Ryabinin, some mathematical results and examples from his book (Ryabinin, 2000) are reproduced. The technology of the automated construction and the analysis of LP-models of any complexity are presented following works by Mojaevand Gromov(2000).

Since a correct and timely decision can have a significant impact on the personal and social life of humans, the need for a strong technique that can help a person in this area is quite tangible. One of the most effective of these techniques is the analytical hierarchy process (AHP), which was first introduced by Thomas L. Saaty in the 1970s. This technique is based on paired comparisons and allows the managers to examine different scenarios. This process has been welcomed by various managers and users in light of its simple yet comprehensive nature until, by comparing the two criteria and sub-criteria in this process, the results of this method will be closer to the actual reality. Based on this, considering that any criterion or sub-criterion in this process has different utility at different levels, it is best to compare them with two criteria according to the desirability of the criteria at each level. To test the results of this work, the technique is used to solve a problem that is available in this book.

The world around us is fraught with multi-criteria issues, and people are always forced to make decisions in these areas. For example, when choosing a job, there are various criteria, such as social status, creativity and innovation, and so on. The decision maker must consider the various options according to these criteria. In large-scale decisions such as annual budget planning, experts have pursued various goals, such as security, industrial development training, etc., and would like to optimize these goals. In the life of the day, there are many examples of decision making with multiple criteria.

In some cases, the result of decision making is critical to the fact that an error may impose irreparable losses on us. Therefore, it is necessary to design the appropriate technique or techniques for optimal selection and decision making so that the decision maker can make the best possible selection closer. The AHP method, based on human brain analysis for complex and fuzzy problems, was suggested by a researcher named Thomas L. Saaty in the 1970s. The hierarchical analysis process is one of the most comprehensive systems designed for decision making with multiple criteria because this technique allows formulation of the problem in a hierarchical manner, as well as the possibility of considering different quantitative and qualitative criteria in the problem is that this process interferes with different options in decision making and allows sensitivity analysis to be based on criteria and sub-criteria, in addition to being based on a paired comparison that facilitates judgment and computation, as well as the degree of adaptability And the incompatibility of the decision demonstrates the privileged advantages of this technique in making a few decisions it helps. The type of our paired comparison between the criteria and the sub-criteria is linear. For example, if the element A preference for element B is always equal to n, the element B preference for element A will always be equal to 1/n, while at the various levels of element A, the element's desirability B has changed. In this research, we have tried to make a more accurate comparison of the criteria and sub-criteria according to the utility theory, which is one of the most applicable theories in microeconomics, and the relative weight of each criterion with the use of the utility function is obtained between the two criteria.

The concept of "optimization" is commonly used both in rational decision theory and natural selection theory. While researchers broadly recognize the differences between the two theories, the differences are not widely emphasized. This chapter aims to stress the differences between the two theories, which warrant calling each concept by a different name: rationality optimization and selection optimization.

The term "rationality" connotes the discipline of economics, while the term "natural selection" connotes the discipline of evolutionary biology. The disciplinary boundary, though, between economics and evolutionary biology is irrelevant to the fault line that separates rationality optimization from selection optimization (Khalil, 2000). Biologists use, without explicitly stating so, the concept of rationality when they discuss the fitness of behavior. On the other hand, economists use, also without explicitly stating so, natural selection when they discuss market equilibrium. So, we need not make a comparison between economics and biology as disciplines, which would usually imply that they use separate conceptual frameworks. In any case, such comparisons have been undertaken (Hirshleifer, 1977; Hodgson, 2004, 2007). The focus of this chapter is, rather, on the concept of optimization and, in particular, how far apart rationality optimization is from selection optimization.

It is imperative to emphasize the difference between rationality optimization and selection optimization. Once the difference is clarified, it is not easy any longer for natural selection theory to explain the origin of rationality optimization (Khalil, 2007b). The basic difference is that natural selection operates at the level of the population as an unintended outcome, while rational decision operates at the level of the individual as an intended action.

In stressing the difference between rationality optimization and selection optimization, this chapter emphasizes the role of behavior and, hence, the development of the organism (ontogeny), which is not the same as the operation of natural selection. The neo-Darwinian theory of evolution, i.e. natural selection theory, has long ignored the role of ontogeny, which is greatly influenced by decision making. However, a new emphasis on the role of ontogeny in evolution has spearheaded by the research program known as Evo Devo (Müller and Newman, 2003). As indicated below, evolutionary economics associated with the work of Joseph Schumpeter, which stresses the learning and development of the firm, parallels this Evo-Devo approach in biology.

One payoff of highlighting the difference between the two kinds of optimization is showing that biologists have been employing the tools of rationality optimization in their analysis of animal behavior and ontogeny without being aware of doing so. Rationality optimization, which amounts to responsiveness to incentives or constraints, is nothing other than what biologists call "phenotypic plasticity." Biologists widely recognize phenotypic plasticity across all taxa and kingdoms. That is, rational decision making typifies the behavior of all organisms – viz., from plants to fungi and animals (Khalil, 2007a).

1.2 Optimization in economics and evolutionary biology

The term "optimization" denotes maximization in natural selection theory and rationality theory. There is a deep difference, though, between the concepts of optimization in both theories. For natural selection theory, optimization is about maximizing the frequency of the fittest trait/technology type in a population. For rationality theory, on the other hand, optimization is about maximizing the objective function of the agent.

This difference does not run along the fault that supposedly separates the sciences that study nonhuman behavior from the sciences that study human behavior. This chapter does not recognize such a fault because, as Figure 1.1 shows, biologists also use rationality optimization, while economists use natural selection optimization. The two concepts of optimization cut across the disciplinary divide. Both selection theory and rationality theory are present in the disciplines of economics and biology. This is the case, irrespective of whether one regards, from the perspective of selection optimization, competitive forces as exact and sharp ("unbounded") or inexact and soft ("bounded") or, from the perspective

Optimization criterion according to.

	Rationality Theories	**Selection Theories**
Economics	Maximization of surplus/utility; Theory of altruism	Market selection of best technology and institutions
Biology	Optimum foraging theory; Inclusive fitness	Natural selection of best genotype

Figure 1.1 Optimization economics and biology.

of rationality optimization, rational decision as exact and sharp (unbounded) or inexact and soft (bounded) (Khalil, 2007a, 2007b).

It is the practice of economists to use rationality optimization. It is less known that they also use selection optimization, but do so under different terminology. Likewise, it is the practice of evolutionary biologists to use selection optimization. It is also less known that they too use rationality optimization, but usually disguise it as "as-if" arguments, i.e. as the study of behavior as if the organisms were rational.

Whenever economists invoke forces of competition to explain why certain technologies survive and become dominant over less adaptive technologies, they are using natural selection theory.[2] Even though there is no sexual or asexual replication in market competition, there is imitation. That is, imitation ensures replication. Imitation does not have to be intentional. The selectionsargument works even if we assume that imitation is random, as agents who are more successful (rational) are more likely to be copied. This is because the more successful type is bound to grow, via investment or consumption, and hence is bound to become more prevalent or noticeable. It is more likely for other agents to imitate the more noticeable trait than imitate the less noticeable treat. So, imitation leads to the dominance of the most productive type. In this manner, imitation in social theory is analogous to reproduction in natural selection theory. So, the selection mechanism in nature or society can generally be called "selection optimization."

It is important to note the two notions of optimization implied here – being the most successful type and the proliferation of the successful type – and to keep them separate. To use the same term to denote both notions of optimization is at the origin of the common accusation that neo-Darwinian theory is "tautological."[3]

A neo-Darwinian theory will amount to a tautology if it postulates that the dominant lineage (the most frequent in the population) must also be the fittest and *vice versa*. Once the two concepts of optimization are separated; however, it is easier to see that the determination of the fittest differs from the determination of the dominant lineage. The actual dominant lineage may not reflect the fittest trait in the case when competition or selection optimization is bounded in the sense of being imperfect. Selection optimization, which leads to equilibrium, is bounded when it allows for unfit firms or unfit lineages to survive and reproduce in the long term. If so, whatever is dominant (determined by selection optimization) does not reflect what the fittest (determined by rationality optimization) is.

The issue of whether selection optimization is bounded or unbounded is an empirical question (see Khalil, 2007a, 2007b). The point is that we have to be careful when we use the term "optimization." Does it denote the fittest/rational decision, or does it denote the dominance of the fittest lineage in the population? Although the two are the same when selection optimization or market forces are unbounded, it is still important to avoid the conflation of the two kinds of optimization.

The study of behavior especially behooves us to keep the two kinds of optimization apart. Biologists generally do not keep them apart. When they employ rationality optimization, they view it as a form of selection optimization, and use terms such as "making decisions" and "rational thinking" under the notion "as-if." That is, they show that organisms think and behave rationally, but only "as-if" as a shorthand that they are dictated to do so as a result of selection optimization. The last section of the chapter illustrates this claim, posed in the lower-left cell in Figure 1.1, by specifically analyzing the inclusive fitness hypothesis, which is usually advanced to explain altruistic behavior.

It may be of no surprise to many that optimum foraging theory is classified as a theory of rationality optimization. Ethologists and ecologists are usually more explicit in the use of the rationality optimization approach when they discuss how organisms tend to act efficiently (MacArthur and Pianka, 1966; Charnov, 1976; McFarland, 1977). Optimum foraging theorists borrowed their tools consciously from rational choice theory in economics. Optimum foraging theory, to which some economists have contributed (Tullock, 1971), is about the organism's search/feeding behavior (Schoener, 1971; Winterhalder and Smith, 1981; Stephens and Krebs, 1986; Smith in Dupré, 1987; Winterhalder, 1992, 2000; Smith and Winterhalder, 1992). The theory takes the organism's ability to extract information from the environment (such as the length of day, temperature, the location of prey, and so on) as given by the genetic trait or genetic type of the organism, and then analyzes the effective use of the organism's capacity type. According to the theory, animals, plants, and other organisms choose the least costly method of producing or catching a given prey, nutrients, or sunlight, so that the harvest per unit of time/effort is maximized. So, optimum foraging theory in biology is identical to rationality optimization theory in economics. In both cases, the agent tries to optimize the objective function, given its input endowment and capacity type.

Likewise, theories of nonhuman societies rely heavily on rational models of division of labor. For instance, the division of labor in ant societies exhibits great phenotypic plasticity, i.e. changes of phenotype (including behavior) without corresponding changes in genotype. Also, recent evidence shows that some animals, including, specifically speaking, the western scrub-jay (Aphelocoma californica), do plan for the future by storing food in response to constraints (Raby et al., 2007). This has prompted some biologists to adopt explicit rational choice models (Cassill, 2003; Vermeij, 2004; Hurley and Nudds, 2006; cf. Mokyr, 2006). Once again, even some economists have contributed to this literature (Landa, 1986; Landa and Wallis, 1988). Another line of research spearheaded by economists (Kagel et al., 1995) has shown how animals in laboratories, such as pigeons and rats, economize in their choices. This confirms the predictions of rationality optimization theory.

Given that economics and biology have used both concepts of optimization, they tend to treat rationality as a product of selection, and hence overlook their differences. To a great extent, there is a redundancy between the operation and function of the two kinds of optimization.

1.3 Explaining behavior

Many practitioners do not limit, though, The Redundancy Thesis, to formal identity. They tend to regard the formal identity as also implying similarity of explanation of behavior. This forces us to investigate whether the two types of optimization are materially identical or not.

The most important feature of neo-Darwinian theory (Mayr, 1976, 1982, 1988; Dawkins, 1976, 1982; Brandon and Burian, 1984; Bonner, 1988) is that agents do not change as a result of selection optimization. Rather, it is the population that undergoes improvement through "editing," i.e. deleting relatively less-fit members. In contrast, the most important feature of rationality theory is that the agent itself changes, by changing its behavior in response to incentives, as a result of rationality optimization. What exactly, then, is the difference between the conceptual apparatuses of the two kinds of optimization? To answer this question, we first need to distinguish two major criteria, the "unit" and the "objective function," as shown in Figure 1.2, about each kind of optimization.[4]

Concerning the horizontal axis of Figure 1.2, the unit criterion specifies what the objective function is made. For rationality optimization, the unit is the agent or any decision maker that can range from the cell, organism, colony of organisms, to the social pack like human organizations (tribes, firms, and states). The unit-as-agent makes decisions to satisfy its needs, given its traits and environmental constraints. For selection optimization, on the other hand, the unit is the population. The population cannot act as an agent because it is not a decision maker. The population is a collection of agents delimited from other populations by the fact that its agents are involved in the competition since they rely on the same ecological niche for their survival. Thus, the population is not even a colony, such as an insect colony, because it does not possess hierarchy, organization,

	Optimization according to.	
	Rationality Theories	**Selection Theories**
Unit	Agent	Population
Objective Function	Utility or Output	The frequency of trait/pattern

Figure 1.2 Optimization rationality vs. selection.

or leadership. The population is instead simply a collection or organisms that happen to compete with each other. The frequency of traits in the population is a product of natural selection, as nature selects differentially among different trait lineages. The population can be considered as a unit only insofar as it is the by-product of natural selection, which is operating at the level of agents that make up the population.

At first approximation, we can assume that the forces of selection operate as if each member's type has no externality effect on the payoffs for other members of the same type. That is, natural selection operates directly on individuals in disregard of the frequency of a given trait in the population, even though the members of the group may not be isolated individuals, and that the payoff derived from their trait or type might be dependent on this frequency. Such frequency-dependent fitness payoffs are the basis of evolutionary game theory (Weibull, 1995) and new attempts to conceive the population as a coherent society (see Millstein, 2006). The fitness of, for example, a poisonous trait that deters predators is a positive function of the frequency of such a trait in the population. If the trait is below a critical level, predators would take the risk and pursue the prey.

Even if the population is not only a collection of agents but rather a coherent society, such as human society, where the fitness of certain lineages, such as trustworthiness (seeBowles, 2004, ch. 2; Gintis et al., 2005), has frequency-dependent payoffs, the population is still not an agent. Society is not a decision maker until it becomes a formal organization. For instance, a society of sovereign and independent tribes would no longer be a unit-as-population, and would rather be a unit-as-agent, only when the tribes become united into a single political authority. The population, in this case, will thus be the totality of this new unit-as-agent and other similar agents in competition with each other. The fact that the fitness of a lineage can be frequency-dependent does not qualify the population as being an agent.

Thus, the issue of frequency-dependent fitness payoffs should not concern the definition of population-as-unit. The unit in natural selection theory is thus the population, in the sense of the by-product of selection operating at the level of agents whether the lineage is frequency-dependent or frequency-independent.

As a unit of evolution, the population becomes "adapted" to its environment in a blind manner. Such adaptation is called here "selection optimization." Put differently, the population is not a decision making unit or, in Richard Dawkins's (1976) lexicon, a vehicle even when the members of the population share the same taxonomic framework, or share, as what Dawkins calls, the same "replicator" (see Khalil, 1997). A population in the neo-Darwinian sense is a unit consisting of members that compete for reproduction and resources. An evolutionary change occurs when the members, because of some exogenous shock, can be ranked according to differential levels of reproduction success, a ranking that the selecting mechanism (nature or the market) can distinguish. A change that a population can experience is thus only merely an unintentional product of the deletion of the less fit trait/pattern types, as measured by the infrequency of their offspring. Such a change is not occasioned by an active decision, as in the case with the change that one experiences when one acts rationally.

Moving to the vertical axis of Figure 1.2, it should also be noted that the "objective function" is also different for both kinds of optimization. For rationality optimization, the function is either the agent's utility or the agent's output or offspring. For selection optimization, the function is the frequency of the fittest trait/pattern type in a population. The selector (nature or the market) "makes" the population of organisms or firms dominated by the best trait/pattern type possible vis-à-vis the environment.[5]

With rationality optimization, each agent maximizes its objective function. This does not need to assume heterogeneity of agents. However, for selection optimization, the assumption of initial heterogeneity of agents is essential to explain the change. Let us take, e.g., a population of 100 agents. If we assume total heterogeneity with regard only to a single lineage, while all other lineages are identical, the population would consist of 100 lineages from which nature can select. Nature then selects the fittest lineage out of the existing 100 lineages. Selection optimization, over time, makes (if optimization is unbounded) the fittest lineage the most frequent and, under some conditions, the only lineage in the population. So, in equilibrium, selection optimization (if unbounded and not facing obstacles similar to what economists call transaction costs) should lead either too minimal heterogeneity or no heterogeneity at all. Put differently, while selection optimization starts with a variety of lineages, as expressed in the population, the selection produces, at first approximation, a population that is almost or homogeneous.

The starting point of selection optimization, variety or heterogeneity, is the product of mutations that occur as a result of exogenous shocks. These mutations take place without any regard to ex-post benefit. Rather, selection optimization makes the best out of what exists in terms of lineages and environmental constraints by making the fittest lineage the dominant one in the population.

This is mathematically similar to rationality optimization, viz., it assures the greatest benefit from using the given resources.

Both kinds of optimization are mechanical in that they do not leave room for endogenous mutation. That is, they do not allow the organism to come up with a mutation in light of ex-post benefit. To state this more specifically in respect to human action, rationality optimization is mechanical in the sense that it does not allow the agent to introduce an innovation in light of its expected benefits. Rationality optimization only specifies the allocation of resources, given the existing menu of innovations. In other words, rationality optimization does not allow for creativity, entrepreneurship, and imagination of the familiar building blocks of Schumpeter's theory of evolutionary change. For rationality optimization, technological innovation is thus ultimately the result of an exogenous process. Aside from the mathematical and mechanical similarities, however, the two kinds of optimization involve different conceptual apparatuses. Biologists seem to recognize the difference between the two kinds of optimization when they label the function that the organism optimizes as the "fitness function," while labeling the function that nature optimizes as the "adaptive or superior fitness function."To recapitulate the argument, selection optimization involves, if the optimization is unbounded, market equilibrium or lineage frequency equilibrium, with which there is no long-term diversity of payoffs across agents belonging to the same competitive population. Of course, the equilibrium would change as a result of a change in environmental constraints (the selectors). Such a change arising from selection optimization differs, however, from an adjustment arising from rationality optimization, i.e. a change in the choice of organisms, from plants to amoeba and humans, in response to a change in incentives.

We have two distinct kinds of optimization: rationality optimization and selection optimization (or natural selection and market equilibrium). Irrespective of the origin of rationality, the objective function that rationality optimizes differs from the function that selection optimizes. Thus, rationality optimization and selection should not be assumed to be identical, even given the fact that both kinds of optimization, under the same assumptions, necessarily engender the same allocation of resources.

> Given that rationality and selection are different, what, then, is the source of rationality? This is an important question, especially if it is true that selection optimization cannot explain rationality optimization (see Khalil, 2008a, b). However, it is also a difficult question that many may try to avoid by hiding under the skirts of Neo-Darwinism. However, this should not discourage us from attempting to answer an easier, but a related, question. Which of the two optimizations is the better methodology to adopt, given that they both engender the same optimum allocation of resources and can, therefore, be used to deter- mine the same answer? If it is the case that selection optimization cannot explain rationality optimization, it behooves us to adopt rationality optimization. However, is it indeed truly the case that

natural selection cannot account for rationality? This question is outside the scope of this paper. However, we would not be able even to ask such a question if we fail to identify the difference between rationality optimization and selection optimization in the first place.

Notes

1 We start to consider the logic and probabilistic (LP) risk LP-theory for systems with groups of incompatible events (GIE); elements and output characteristics of these systems have some levels of values. Examples of such systems are: credits in banks, security portfolio, diagnosing and monitoring, estimation of quality of company work, accuracy of production, etc., in which the risk is the usual and mass phenomenon, and there are sufficient statistical data on risk objects (Solojentsev et al., 1999; Solojentsevand Alekseev, 2003).

2 In fact, the extension of natural selection arguments is not limited to economics. A growing number of social scientists are attempting to reformulate their respective disciplines according to evolutionary biology. This is especially evident in psychology (Barkow et al., 1992), anthropology (Boyd and Richerson, 1985; Betzig et al., 1988), sociology (Machalek, 1992; Lopreato and Crippen, 1999), and political science (Masters and Gruter, 1992). In economics, this "invasion" has gone in diverse directions (Hirshleifer, 1982; Nelson and Winter, 1982; Anderson et al., 1988; Hodgson, 1993, 2002; Nelson, 1995; passim Koslowski, 1999; Witt, 2003). On the other hand, a few thinkers even aspire to ground the first principles of biology on cost-benefit analysis (Ghiselin, 1974, 1992; Tullock, 1994). And they are not alone. The study of animal behavior, such as the behavior of specific social insects, is based greatly on how agents exchange information and adjust behavior in light of costs and benefits (Detrain et al., 1999; Cassill, 2003; Franks et al., 2003). It can even be concluded that the aspiration of a general theory of behavior is not unreasonable (see Knudsen, 2002).

3 We ignore here many of the nuances of the term "fitness." In particular, we ignore the issue of actual fitness as opposed to expected fitness (see Endler, 1986, pp. 27–51). Such details are unrelated to our main argument. It should also be noted that there is a slight ambiguity in the literature concerning the definition of fitness for sexually reproducing organisms (Keller, 1987). The measure of fitness in terms of the quantity of individuals born with the robust type differs from the number of individuals to which each fit agent gives birth for the simple reason that it takes two agents to replicate in sexual reproduction. Furthermore, a more important problem, which is overlooked here, is that natural selection in a sexually reproducing population may not necessarily engender fitness (Akin, 1979; Karlin and Lessard, 1986). The selected differences at the phenotypic level may not be transmitted to the next generation because of the random reshuffling of genes, which is responsible for the probabilistic character of Mendelian inheritance.

4 As mentioned below, Sober (1998) also highlights the difference between the two kinds of optimization. Sober, though, discusses a third criterion, besides the two mentioned in the text, which sets rationality optimization apart from selection optimization. Mindless organisms supposedly do not have subjective utility, but they still have the objective property of fitness. However, as shown earlier, utility optimization, which is used to characterize human decision making, parallels foraging optimization; and market selection parallels natural selection. Thus, although utility can be maximized by agents using anything that they regard as conducive to their welfare, utility itself is still an objective property. Thus, Sober's third distinction is unwarranted.

5 In the sense used here, selection should be distinguished from the account of the rise of conventions or standards which are usually welfare- or fitness-neutral vis-à-vis the environment. The stability of conventions – such as using the metric system or particular facial expressions to express disapproval – depends on what other members of the group are doing (Young, 1996). In biology, the theory of "evolutionarily stable strategy" (ESS) and evolutionary game theories provide, inter alia, an account of conventions (Maynard Smith, 1978a, 1978b, 1982; Vincent and Brown, 1988; Hammerstein and Selten, 1994). A strategy is found to be ESS if all members in the pertinent population adopt it, which makes the group immune from the invasion of other competing strategies. In contrast, the stability characterizing the fitness of a population in relation to its environment is a substantive property, i.e. not conditioned on the unison of actions of members. This paper is concerned exclusively with substantive properties which are usually welfare- or fitness-sensitive vis-à-vis the environment.

2 Risks in economics

2.1 Utility theory

The concept of utility and its relationship with the value of goods and services.

Consumer satisfaction, which results from the consumption of goods and services, is called utility by economists.

Professor Jermain Bentham of English first introduced the term utility. Nevertheless, neither her nor the economists of her age understood the relationship between the value of goods and the utility they derive from the consumption of goods.

Adam Smith recognized the relationship between the value of the use and the value of the commodity's exchange and described his famous example of water and diamonds. Diamonds have a high price, but its value for life is low, while water prices (exchange value) are small but essential for life, and they have a very high value.

2.1.1 Cardinal Theory of Utility

The usefulness of cardinal theory suggests that utility is measured just as much as the price and amount. This means that we can determine the amount of utility of each commodity. In this theory, both utility and ultimate utility can be measured. Economists who believed that cardinality was desirable can be divided into two groups:

1 Those who believed that the utility was acceptable.
2 Those who believed that utility could not be comprehensible.

To better understand the subject of this theory, consider the following article that is provided by J. Singh.

2.1.2 Consumer's Behavior. Cardinal Utility Analysis (Explained with Diagram)

From time to time, different theories have been advanced to explain consumer's demand for a good and to derive a valid demand theorem.

Cardinal utility analysis is the oldest theory of demand which provides an explanation of consumer's demand for a product and derives the law of demand which establishes an inverse relationship between price and quantity demanded of a product.

The price of a product depends upon the demand for and the supply of it. In this part of the book, we are concerned with the theory of consumer's behavior, which explains his demand for a good and the factors determining it. Individual's demand for a product depends upon the price of the product, the income of the individual and the prices of related goods.

It can be put in the following functional form:

$$D_x = f(P_x, I, P_y, P_2, T, \text{etc.}) \tag{2.1}$$

where D_x stands for the demand of good X, P_x for the price of good X, I for individual's income, $P_y \cdot P_z$ for the prices of related goods and T for tastes and preferences of the individual. However, among these determinants of demand, economists single out the price of the good in question as the most important factor governing the demand for it. Indeed, the function of a theory of consumer's behavior is to establish a relationship between quantity demanded of a good and its price and to provide an explanation for it.

Recently, cardinal utility approach to the theory of demand has been subjected to severe criticisms, and as a result, some alternative theories, namely, Indifference Curve Analysis, Samuelson's Revealed Preference Theory and Hicks' Logical Weak Ordering Theory have been propounded.

2.1.3 Assumptions of Cardinal Utility Analysis

Cardinal utility analysis of demand is based upon certain important assumptions. Before explaining how cardinal utility analysis explains consumer's equilibrium regarding the demand for a good, it is essential to describe the basic assumptions on which the whole utility analysis rests. As we shall see later, cardinal utility analysis has been criticized because of its unrealistic assumptions.

The basic assumptions or premises of cardinal utility analysis are as follows.

2.2 The Cardinal Measurability of Utility

The exponents of cardinal utility analysis regard utility to be a cardinal concept. In other words, they hold that utility is a measurable and quantifiable entity. According to them, a person can express utility or satisfaction he derives from the goods in the quantitative cardinal terms. Thus, a person can say that he derives utility equal to 10 units from the consumption of a unit of good A, and 20 units from the consumption of a unit of good B.

Moreover, the cardinal measurement of utility implies that a person can compare utilities derived from goods in respect of size, that is, how much one level of utility is greater than another. A person can say that the utility he gets from the consumption of one unit of good B is double the utility he obtains from the consumption of one unit of good A.

According to Marshall, marginal utility is measurable in terms of money. Money represents the general purchasing power, and it can, therefore, be regarded as command over alternative utility-yielding goods. Marshall argues that the amount of money which a person is prepared to pay for a unit of a good rather than go without it is a measure of the utility he derives from that good.

Thus, according to him, money is the measuring rod of utility. Some economists belonging to the cardinalist school measure utility in imaginary units called "utils." They assume that a consumer is capable of saying that one apple provides him utility equal to 4 utils. Further, on this ground, he can say that he gets twice as much utility from an apple as compared to an orange.

2.3 The Hypothesis of Independent Utilities

The second important tenet of the cardinal utility analysis is the hypothesis of independent utilities. On this hypothesis, the utility which a consumer derives from a good is the function of the quantity of that good and of that good only in other words, the utility which a consumer obtains from a good does not depend upon the quantity consumed of other goods; it depends upon the quantity purchased of that good alone.

On this assumption, then the total utility which a person gets from the whole collection of goods purchased by him is simply the total sum of the separate utilities of the goods. Thus, the cardinalize school regards utility as 'additive,' that is, separate utilities of different goods can be added to obtain the total sum of the utilities of all goods purchased.

2.4 The constancy of the Marginal Utility of Money

Another important assumption of the cardinal utility analysis is the constancy of the marginal utility of money. Thus, while the cardinal utility analysis assumes that marginal utilities of commodities diminish as more of them are purchased or consumed, the marginal utility of money remains constant throughout when the individual is spending money on a good and due to which the amount of money with him varies. Daniel Bernoulli, first of all, introduced this assumption, but later Marshall adopted this in his famous book "Principles of Economics."

As stated above, Marshall measured marginal utilities in terms of money. However, the measurement of the marginal utility of goods in terms of money is only possible if the marginal utility of money itself remains constant. It should be noted that the assumption of constant marginal utility of money is very crucial to the Marshallian analysis because otherwise, Marshall could not measure the marginal utilities of goods in terms of money. If money, which is the unit of measurement itself, varies as one is measuring with it, then it cannot yield a correct measurement of the marginal utility of goods.

When the price of good falls, and as a result, the real income of the consumer rises, the marginal utility of money to him will fall, but Marshall ignored this and assumed that marginal utility of money did not change as a result of the change in price. Likewise, when the price of a good rises, the real income of the consumer will fall, and his marginal utility of money will rise. However, Marshall ignored this and assumed that the marginal utility of money remains the same. Marshall defended this assumption on the ground that "his (the individual consumer's) expenditure on any one thing is only a small part of his whole expenditure."

2.5 Introspective Method

Another important assumption of the cardinal utility analysis is the use of the introspective method in judging the behavior of marginal utility.

> Introspection is the ability of the observer to reconstruct events which go on in the mind of another person with the help of self-observation. This form of comprehension may be just guesswork or intuition or the result of long-lasting experience.

Thus, the economists construct with the help of their own experience the trend of feeling which goes on in other men's mind. From his response to certain forces and from experience and observation, one gains an understanding of the way other people's minds would work in similar situations. To sum up, in the introspective method, we attribute to another person what we know of our mind. That is, by looking into ourselves, we see inside the heads of other individuals.

So, the Law of Diminishing Marginal Utility is based upon introspection. We know from our mind that as we have more of a thing, the less utility we derive from an additional unit of it. We conclude from it that other individuals' minds will work similarly, that is, a marginal utility to them of goodwill diminishes as they have more units of it.

With the above basic premises, the founders of cardinal utility analysis have developed two laws which occu*py* an important place in economic theory and have several applications and uses.

2.5.1 These two laws are

1 Law of Diminishing Marginal Utility and
2 Law of Equi-Marginal Utility.

It is with the help of these two laws about consumer's behavior that the exponents of cardinal utility analysis have derived the law of demand. We explain below these two laws in detail and how the law of demand is derived from them.

2.5.2 Law of Diminishing Marginal Utility

An important tenet of cardinal utility analysis relates to the behavior of marginal utility. This familiar behavior of marginal utility has been stated in the Law of Diminishing Marginal Utility according to which marginal utility of a good diminishes as an individual consumes more units of a good. In other words, as a consumer takes more units of a good, the extra utility or satisfaction that he derives from an extra unit of the good goes on falling.

It should be carefully noted that it is the marginal utility and not the total utility that declines with the increase in the consumption of a good. The Law of Diminishing Marginal Utility means that the total utility increases at a decreasing rate.

Marshall who has been a famous exponent of the cardinal utility analysis has stated the Law of Diminishing Marginal Utility as follows:

The additional benefit which a person derives from a given increase of his stock of a thing diminishes with every increase in the stock that he already has.

This law is basedon two important facts. First, while the total wants of a man are virtually unlimited, every single want is satiable. Therefore, as an individual consumes more and more units of a good, intensity of his want for good goes on falling, and a point is reached where the individual no longer wants any more units of the good. That is when the saturation point is reached, and the marginal utility of a good becomes zero. Zero marginal utility of a good implies that the individual has all that he wants of the good in question.

The second fact on which the Law of Diminishing Marginal Utility is based is that the different goods are not perfect substitutes for each other in the satisfaction of various wants. When an individual consumes more and more units of a good, the intensity of his particular want for the good diminishes but if the units of that good could be devoted to the satisfaction of other wants and yielded as much satisfaction as they did initially in the satisfaction of the first want, marginal utility of the good would not have diminished.

It is obvious above that the Law of Diminishing Marginal Utility describes a familiar and fundamental tendency of human nature. This law has been arrived at by introspection and by observing how consumers behave.

2.5.3 Illustration of the Law of Diminishing Marginal Utility

Consider Table 2.1 where we have presented the total and marginal utilities derived by a person from cups of tea consumed per day. When one cup of tea is taken per day, the total utility derived by the person is 12 utils. Moreover, because this is the first cup, its marginal utility is also 12 utils with the consumption of second cup per day; the total utility rises to 22 utils, but marginal utility falls to 10. It will be seen from the table that as the consumption of tea increases to six cups per day, marginal utility from the additional cup goes on diminishing (i.e. the total utility goes on increasing at a diminishing rate).

However, when the cups of tea consumed per day increase to seven, then instead of giving positive marginal utility, the seventh cup gives negative marginal utility equal to – 2 utils. This is because too many cups of tea consumed per day

Table 2.1 Diminishing marginal utility

Cups of tea consumed per day (Q)	*Total utility (Utils) (TU)*	*Marginals utility (Utils)* $\frac{\Delta TU}{\Delta Q}$
1	12	12
2	22	10
3	30	8
4	36	6
5	40	4
6	41	1
7	39	-2
8	34	-5

(say more than six for a particular individual) may cause acidity and gas trouble. Thus, the extra cups of tea beyond six to the individual in question give him disutility rather than positive satisfaction.

Figure 2.1 illustrates the total utility and marginal utility curves. The total utility curve drawn in Figure 2.1 is based upon three assumptions. First, as the quantity consumed per period by a consumer increases, his total utility increases but at a decreasing rate; this implies that as the consumption per period of a commodity by the consumer increases, marginal utility diminishes as shown in the lower panel of Figure 2.1.

Second, as will be observed from the figure when the rate of consumption of a commodity per period increases to Q_4, the total utility of the consumer reaches its maximum level.

Therefore, the quantity Q_4 of the commodity is called satiation quantity or satiety point. Third, the increase in the quantity consumed of the good per period by the consumer beyond the satiation point hurts his total utility that is, his total utility declines if more than Q_4 quantity of the good is consumed.

This means that beyond Q_4, marginal utility of the commodity for the consumer becomes negative, and ads will be seen in the lower panel of Figure 2.1; beyond the satiation point Q_4, marginal utility curve MU goes below the *X*-axis, indicating that it becomes negative beyond the quantity Q_4 per period of the commodity consumed.

It is important to understand how we have drawn the marginal utility curve. As stated above, marginal utility is the increase in total utility of the consumer caused by the consumption of an additional unit of the commodity per period. We can directly find out the marginal utility of the successive units of the commodity consumed by measuring the additional utility which a consumer obtains from successive units of the commodity and plotting them against their respective quantities.

However, in terms of calculus, the marginal utility of a commodity *X* is the slope of the total utility function

$$U = f(Q_x) \tag{2.2}$$

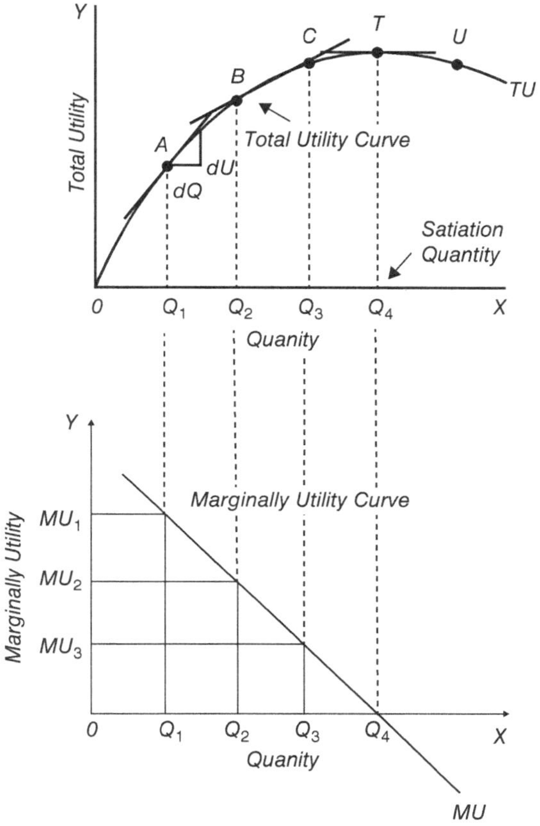

Figure 2.1 Total utility and marginal utility.

Thus, we can derive the marginal utility curve by measuring the slope at various points of the total utility curve TU in the upper panel of Figure 2.1, by drawing tangents at them. For instance, at the quantity Q_1 marginal utility

$$(\text{i.e. } dU/dQ = \text{MU}_1) \tag{2.3}$$

is found out by drawing a tangent at point A and measuring its slope which is then plotted against quantity in the lower panel of Figure 2.1. In the lower panel, we measure marginal utility of the commodity on the *Y*-axis. Likewise, at the quantity Q_2, marginal utility of the commodity has been obtained by measuring the slope of the total utility curve TU at point B and plotting it in the lower panel against the quantity Q_2.

It will be seen from the figure that at Q_4 of the commodity consumed, the total utility reaches at the maximum level *T*. Therefore, at the quantity Q_4, the slope of the total utility curve is zero at this point. Beyond the quantity Q_4, the total utility declines and marginal utility becomes negative. Thus, the quantity Q_4 of the commodity represents the satiation quantity.

Another important relationship between total utility and marginal utility is worth noting. At any quantity of a commodity consumed, the total utility is the sum of the marginal utilities. For example, if the marginal utilities of the first, second and third units of the commodity consumed are 15, 12 and 8 units, then the total utility obtained from these three units of consumption of the commodity must equal 35 units ($15 + 12 + 8 = 35$).

Similarly, in terms of graphs of total utility and marginal utility depicted in Figure 2.1, the total utility of the quantity Q_4 of the commodity consumed is the sum of the marginal utilities of the units of the commodity up to point Q_4. That is, the entire area under the marginal utility curve MU in the lower panel up to a pointQ_4 is the sum of marginal utilities which must be equal to the total utility Q_4Tin the upper panel.

2.6 Marginal Utility and Consumer's Tastes and Preferences

The utility people derive from consuming a particular commodity depends on their tastes and preferences. Some consumers like oranges, others prefer apples and still others prefer bananas for consumption. Therefore, the utility which different individuals get from these various fruits depends on their tastes and preferences.

An individual would have different marginal utility curves for different commodities depending on his tastes and preferences. Thus, the utility which people derive from various goods reflects their tastes and preferences for them. However, it is worth noting that we cannot compare utility across consumers. Each consumer has a unique subjective utility scale. In the context of cardinal utility analysis, a change in consumer's tastes and preferences means a shift in his one or more marginal utility curves.

However, it may be noted that a consumer's tastes and preferences do not frequently change, as these are determined by his habits. Of course, tastes and preferences can change occasionally. Therefore, in economic theory, we generally assume that tastes or preferences are given and relatively stable.

2.7 The significance of Diminishing Marginal Utility

The significance of the diminishing marginal utility of a good for the theory of demand is that it helps us to show that the quantity demanded of a good increase as its price falls and vice versa. Thus, it is because of the diminishing marginal utility that the demand curve slopes downward. If properly understood, the Law of Diminishing Marginal Utility applies to all objects of desire including money.

However, it is worth mentioning that the marginal utility of money is generally never zero or negative. Money represents purchasing power over all other goods, that is, a man can satisfy all his material wants if he possesses enough money. Since man's total wants are practically unlimited, therefore, the marginal utility of money to him never falls to zero.

The marginal utility analysis has a good number of uses and applications in both economic theory and policy. The concept of marginal utility is of crucial significance in explaining the determination of the prices of commodities. The discovery of the concept of marginal utility has helped us to explain the paradox of value which troubled Adam Smith in "The Wealth of Nations."

Adam Smith was greatly surprised to know why water which is so very essential and useful to life has such a low price (indeed no price), while diamonds, which are quite unnecessary, have such a high price. He could not resolve this water-diamond paradox. However, modern economists can solve it with the aid of the concept of marginal utility.

According to modern economists, the total utility of a commodity does not determine the price of a commodity, and it is the marginal utility which is a crucially important determinant of price. Now, the water is available in abundant quantities so that its relative marginal utility is very low or even zero. Therefore, its price is low or zero. On the other hand, the diamonds are scarce, and therefore their relative marginal utility is quite high, and this is the reason why their prices are high.

Prof. Samuelson explains this paradox of value in the following words:

The more there is of a commodity, the less the relative desirability of its last little unit becomes, even though its total usefulness grows as we get more of the commodity. So, it is obvious why a large amount of water has a low price or why air is free good despite its vast usefulness. The many later units pull down the market value of all units.

Besides, the Marshallian concept of consumer's surplus is based on the principle of diminishing marginal utility.

2.8 Consumer's Equilibrium. The principle of Equi-Marginal Utility

The principle of equi-marginal utility occupies an important place in cardinal utility analysis. It is through this principle that consumer's equilibrium is explained. A consumer has a given income which he has to spend on various goods he wants. Now, the question is how he would allocate his given money income among various goods, that is to say, what would be his equilibrium position in respect of the purchases of the various goods. It may be mentioned here that consumer is assumed to be 'rational,' that is, he carefully calculates utilities and substitutes one good for another to maximize his utility or satisfaction.

Suppose there are only two goods X and Y on which a consumer has to spend a given income. The consumer's behavior will be governed by two factors: first, the marginal utilities of the goods and second, the prices of two goods. Suppose the prices of the goods are givento the consumer.

The Law of Equi-Marginal Utility states that the consumer will distribute his money income between the goods in such a way that the utility derived from the last rupee spent on each good is equal. In other words, the consumer is in equilibrium position when the marginal utility of money expenditure on each good is the same. Now, the marginal utility of money expenditure on a good is equal to the marginal utility of a good divided by the price of the good. In symbols,

$$\mathrm{MU}_m = \mathrm{MU}_x / P_x \tag{2.4}$$

where MU_m is the marginal utility of money expenditure,MU_m is the marginal utility of *X*, and P_x is the price of *X*. The Law of Equi-Marginal Utility can, therefore, be stated thus. The consumer will spend his money income on different goods in such a way that marginal utility of money expenditure on each good is equal. That is, consumer is in equilibrium in respect of the purchases of two goods *X* and *Y* when

$$\mathrm{MU}_x / P_x = \mathrm{MU}_y / P_y \tag{2.5}$$

Now, if MU_x/P_x and MU_y/P_y are not equal, and MU_x/P_x is greater than MU_y/P_y, then the consumer will substitute good *X* for good *Y*. As a result of this substitution, the marginal utility of good *X* will fall, and the marginal utility of good *y* will rise. The consumer will continue substituting good *X* for good *Y* until MU_x/P_x becomes equal to MU_y/P_y. When MU_x/P_x becomes equal to MU_y/P_y, the consumer will be in equilibrium.

However, the equality of MU_x/P_x with MU_y/P_y can be achieved not only at one level but at different levels of expenditure. The question is how far does a consumer go in purchasing the goods he wants. The size of his money income determines this. With a given income and money expenditure, a rupee has a certain utility for him. This utility is the marginal utility of money to him.

Since the Law of Diminishing Marginal Utility also applies to money income, the greater the size of his money income, the smaller the marginal utility of money to him. Now, the consumer will go on purchasing goods until the marginal utility of money expenditure on each good becomes equal to the marginal utility of money to him.

Thus, the consumer will be in equilibrium when the following equation holds good.

$$\mathrm{MU}_x / P_x = \mathrm{MU}_y / P_y = \mathrm{MU}_m \tag{2.6}$$

where MU_m is the marginal utility of money expenditure (that is, the utility of the last rupee spent on each good).

If there are more than two goods on which the consumer is spending his income, the above equation must hold good for all of them. Thus,

$$\mathrm{MU}_x / P_x = \mathrm{MU}_y / P_y = \cdots = \mathrm{MU}_m \tag{2.7}$$

Let us illustrate the ***Law of Equi-Marginal Utility*** *with the aid of an arithmetical table given below.*

Table 2.2 The marginal utility of goods X and Y

Units	MU_x *(Utils)*	MU_y *(Utils)*
1	20	24
2	18	21
3	16	18
4	14	15
5	12	9
6	10	3

Let the prices of goods X and Y be Rs. 2 and 3, respectively. We are reconstructing Table 2.2 by dividing marginal utilities (MU) of X by Rs. 2 and marginal utilities (MU) of Y by Rs. 3, we get Table 2.3.

Table 2.3 The marginal utility of money expenditure

Units	$\frac{MU_x}{P_x}$	$\frac{MU_y}{P_y}$
1	10	8
2	9	7
3	8	6
4	7	5
5	6	3
6	5	1

Suppose a consumer has money income of Rs. 24 to spend on the two goods. It is worth noting that in order to maximize his utility, the consumer will not equate marginal utilities of the goods because the prices of the two goods are different. He will equate the marginal utility of the last rupee (i.e. the marginal utility of money expenditure) spent on these two goods.

In other words, he will equate MU_x/P_x with MU_y/P_y while spending his given money income on the two goods. By looking at Table 2.3, it will become clear that MU_x/P_x is equal to 5 utils when the consumer purchases 6 units of good X and MU_y/P_y is equal to 5 utils when he buys 4 units of good Y. Therefore, the consumer will be in equilibrium when he is buying 6 units of good X,and4 units of good Y and will be spending (Rs. 2 × 6 + Rs. 3 × 4) = Rs. 24 on them that is equal to the consumer's given income. Thus, in the equilibrium position where the consumer maximizes his utility,

$$MU_x/P_x = MU_y/P_y = \cdots = MU_m \tag{2.8}$$

$$10/2 = 15/3 = 5$$

Thus, the marginal utility of the last rupee spent on each of the two goods he purchases is the same, that is, 5 utils.

Consumers' equilibrium is graphically portrayed in Figure 2.2. Since marginal utility curves of goods slope downward, curves depicting MU_x/P_x and MU_y/P_y also slope downward. Thus, when the consumer is buying OH of *X* and OK of *Y*, then

$$MU_x/P_x = MU_y/P_y = \cdots = MU_m \tag{2.9}$$

Therefore, the consumer is in equilibrium when he is buying 6 units of *X* and 4 units of *Y*. No other allocation of money expenditure will yield him greater utility than when he is buying 6 units of commodity *X* and 4 units of commodity *Y*. Suppose the consumer buys one unit less of good *X* and one unit more of good *Y*.

This will lead to a decrease in his total utility. It will be observed from Figure 2.2(a) that the consumption of 5 units instead of 6 units of commodity *X* means a loss in satisfaction equal to the shaded area ABCH, and from Figure 2.2(b), it will be seen that the consumption of 5 units of commodity *Y* instead of 4 units will mean a gain in utility equal to the shaded area KEFL. It will be noticed that with this rearrangement of purchases of the two goods, the loss in utility ABCH exceeds gain in utility KEFL.

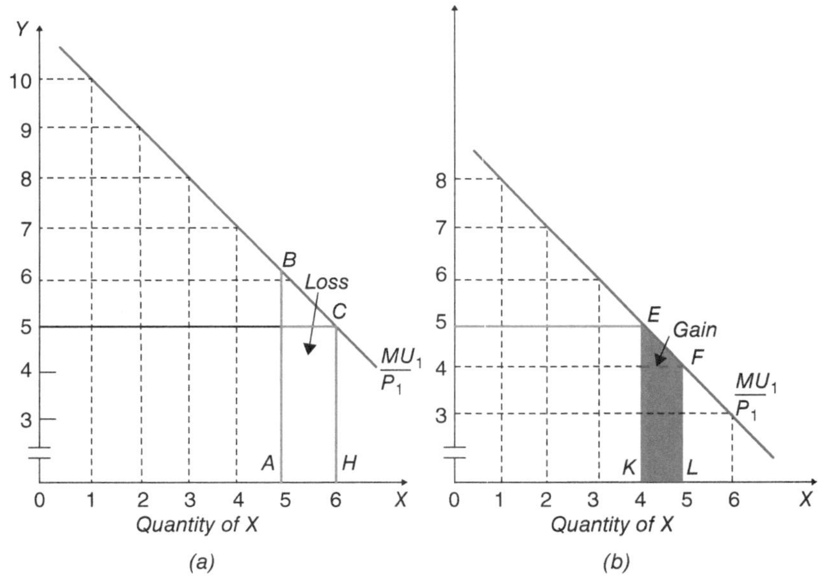

Figure 2.2 Marginal utility principle and consumer's equilibrium.

Thus, his total satisfaction will fall as a result of this rearrangement of purchases. Therefore, when the consumer is making purchases by spending his given income in such a way that

$$MU_x/P_x = MU_y/P_y, \tag{2.10}$$

he will not like to make any further changes in the basket of goods and will, therefore, be in equilibrium situation by maximizing his utility.

2.9 Limitations of the Law of Equi-Marginal Utility

Like other laws of economics, the Law of Equi-Marginal Utility is also subject to various limitations. This law, like other laws of economics, brings out an important tendency among the people. This is not necessary that all people exactly follow this law in the allocation of their money income and therefore all may not obtain maximum satisfaction.

This is due to the following reasons:

1 For applying this Law of Equi-Marginal Utility in real life, the consumer must weigh in his mind the marginal utilities of different commodities. For this, he has to calculate and compare the marginal utilities obtained from different commodities.

   However, it has been pointed out that ordinary consumers are not so rational and calculating. Consumers are generally governed by habits and customs. Because of their habits and customs, they spend particular amounts of money on different commodities, regardless of whether the particular allocation maximizes their satisfaction or not.

2 For applying this law in actual life and equating the marginal utility of the last rupee spent on different commodities, the consumers must be able to measure the marginal utilities of different commodities in cardinal terms. However, this is easier said than done. It has been said that it is not possible for the consumer to measure utility cardinally.

   Being a state of psychological feeling and also there being no objective units with which to measure utility, it is cardinally immeasurable. It is because of the immeasurability of utility in cardinal terms that the consumer's behavior has been explained with the help of ordinal utility by J.R. Hicks and R.G.D. Allen.

3 Another limitation of the Law of Equi-Marginal Utility is found in the case of indivisibility of certain goods. Goods are often available in large indivisible units. Because the goods are indivisible, it is not possible to equate the marginal utility of money spent on them. For instance, in allocating money between the purchase of car and food grains, marginal utilities of the last rupee spent on them cannot be equated.

An ordinary car cost about Rs. 300,000 and is indivisible, whereas food grains are divisible and money spent on them can be easily varied. Therefore, the marginal utility of rupee obtained from cars cannot be equalized with that obtained from food grains. Thus, indivisibility of certain goods is a great obstacle in the way of equalization of marginal utility of a rupee from different commodities.

2.10 Derivation of Demand Curve and the Law of Demand

We now turn to explaining how the demand curve and law of demand are derived in the marginal utility analysis. As stated above, the demand curve or law of demand shows the relationship between the price of a good and its quantity demanded. Marshall derived the demand curves for goods from their utility functions.

It should be further noted that in his utility analysis of demand, Marshall assumed the utility functions of different goods to be independent of each other. In other words, Marshallian technique of deriving demand curves for goods from their utility functions rests on the hypothesis of additive utility functions, that is, utility function of each good consumed by a consumer does not depend on the quantity consumed of any other good.

As has already been noted, in case of independent utilities or additive utility functions, the relations of substitution and Complementarity between goods are ruled out. Further, in deriving demand curve or law of demand, Marshall assumes the marginal utility of money expenditure (Mu_m) in general to remain constant.

We now proceed to derive the demand curve from the Law of Equi-Marginal Utility. Consider the case of a consumer who has a certain given income to spend on some goods. According to the Law of Equi-Marginal Utility, the consumer is in equilibrium regarding his purchases of various goods when marginal utilities of the goods are proportional to their prices.

Thus, the consumer is in equilibrium when he is buying the quantities of the two goods in such a way that satisfies the following proportionality rule:

$$\mathrm{MU}_x/P_x = \mathrm{MU}_y/P_y = \cdots = \mathrm{MU}_m \tag{2.11}$$

where MU_m stands for marginal utility of money income in general.

With a certain given income for money expenditure, the consumer would have a certain marginal utility of money (Mu_m) in general. In order to attain the equilibrium position, according to the above proportionality rule, the consumer will equalize his marginal utility of money (expenditure) with the ratio of the marginal utility and the price of each commodity he buys.

It follows therefore that a rational consumer will equalize the marginal utility of money (MU_m) with MU_x/P_x of good X, with MU_m/P_Y of good Y and so on. Given the Ceteris Paribus assumption, suppose the price of good X falls. With the fall in the price of good X, the price of good Y, consumer's income and tastes remaining unchanged, the equality of MU_x/P_x with MU_y/P_y and MU_m, in general, would be disturbed.

With a lower price than before, MU_x/P_x will be greater than MU_y/P_y or MU_m (it is assumed, of course, that the marginal utility of money does not change as a result of the change in the price of one good). Then, in order to restore equality, the marginal utility of *X* or MU_x must be reduced. Moreover, the marginal utility of *X* or MU_x can be reduced only by the consumer buying more of the good *X*.

It is thus clear from the proportionality rule that as the price of good falls, its quantity demanded will rise, other things remaining the same. This will make the demand curve for a good downward sloping. How the quantity purchased of a good increase with the fall in its price and also how the demand curve is derived in the cardinal utility analysis is illustrated in Figure 2.3.

In the upper portion of Figure 2.3, on the *Y*-axis, MU_x/P_x is shown, and on the *X*-axis, the quantity demanded of good *X* is shown. Given a certain income of the consumer, the marginal utility of money in general for him is equal to OH. The consumer is buying Oq_1 of good *X* when the price is P_{x1} since at the quantity Oq_1 of *X*, the marginal utility of money OH is equal to MU_x/P_{x1}.

Now, when the price of good *X* falls to P_{x2}, the curve will shift upward to the new position MU_x/P_{x2}. In order to equate the marginal utility of money (OH) with the new MU_x/P_{x2}, the consumer increases the quantity demanded to Oq_2. Thus, with the fall in the price of good *X* to P_{x2}, the consumer buys more of it.

It should be noted that no account is taken of the increase in real income of the consumer as a result of the fall in the price of good *X*. This is because if change in real income is taken into account, then marginal utility of money will also change and this would have an effect on the purchases of goods.The marginal utility of money can remain constant in two cases. First, the elasticity of marginal utility curve (price elasticity of demand) is unity so that even with the increase in the purchase of a commodity following the fall in price, the money expenditure made on it remains the same.

Second, the marginal utility of money will remain approximately constant for small changes in the price of unimportant goods, that is, goods which account for negligible part of the consumer's budget. In case of these unimportant goods, increase in real income following the fall in price is negligible and therefore can be ignored.

At the bottom of Figure 2.3, the demand curve for *X* is derived. In the lower panel, the priceis measured on the *Y*-axis. As in the upper panel, the *X*-axis represents quantity. When the price of good *X* is P_{x1}, the relevant curve of MU/*P* is MU_x/P_{x1}, which is shown in the upper panel. With MU_x/P_{x1}, he buys Oq_1 of good *X*. Now, in the lower panel, this quantity Oq_1 is directly shown to be demanded at the price P_{x2}.

When the price of *X* falls to P_{x2}, the curve of MU/P shifts upward to the new position MU_x/P_{x2}. With MU_x/P_{x2}, the consumer buys Oq_2 of *X*. This quantity Oq_2 is directly shown to be demandedat a priceP_{x2} in the lower panel. Similarly, by varying the price further, we can know the quantity demanded at other prices. Thus, by joining points A, B and C, we obtain the demand curve DD. The demand curve DD slopes downward which shows that as the price of good falls, its quantity purchased rises.

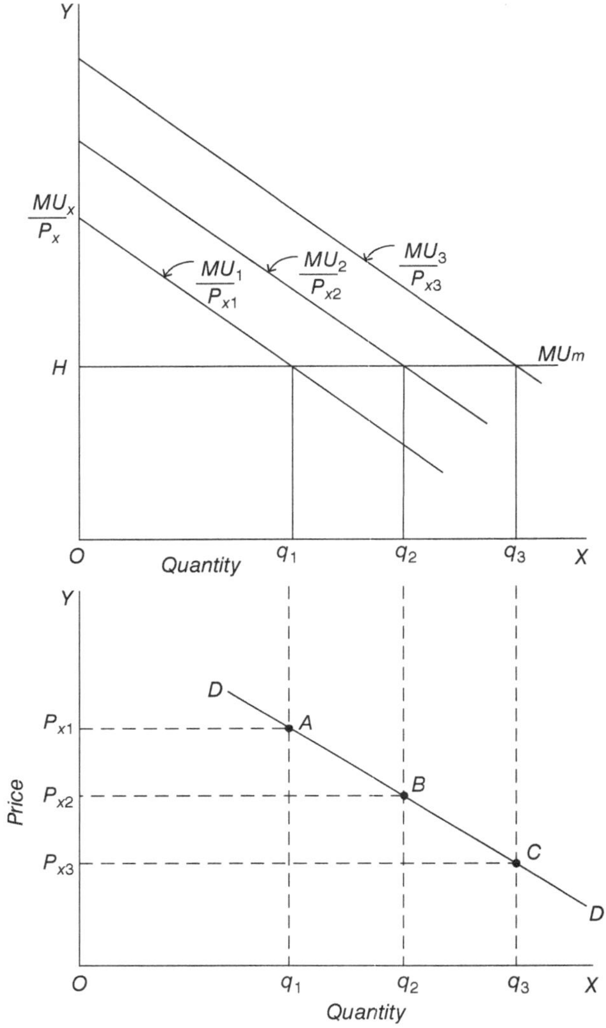

Figure 2.3 Derivation of the demand curve.

2.11 Critical Evaluation of Marshall's Cardinal Utility Analysis

Cardinal utility analysis of demand which we have studied above has been criticized on various grounds.

The following shortcomings and drawbacks of cardinal utility analysis have been pointed out.

2.11.1 Cardinal measurability of utility is unrealistic

Cardinal utility analysis of demand is based on the assumption that utility can be measured in absolute, objective and quantitative terms. In other words, it is assumed in this analysis that utility is cardinally measurable. According to this, how much utility a consumer obtains from goods can be expressed or stated in cardinal numbers such as 1, 2, 3, 4 and so forth. However, in actual practice, utility cannot be measured in such quantitative or cardinal terms.

Since utility is a psychic feeling and a subjective thing, it cannot be measured in quantitative terms. In real life, consumers are only able to compare the satisfaction derived from various goods or various combinations of goods. In other words, in real life, the consumer can only state whether a good or a combination of goods gives him more, less or equal satisfaction as compared to another. Thus, economists like J.R. Hicks think that the assumption of cardinal measurability of utility is unrealistic and therefore it should be given up.

2.11.2 The hypothesis of independent utilities is wrong

Utility analysis also assumes that utilities derived from various goods are independent. This means that the utility which a consumer derives from good is the function of the quantity of that good and of that good alone. In other words, the assumption of independent utilities implies that the utility which a consumer obtains from a good does not depend upon the quantity consumed of other goods; it depends upon the quantity purchased of that good alone.

On this assumption, the total utility which a person gets from the whole collection of goods purchased by him is simply the total sum of the separate utilities of various goods. In other words, utility functions are additive.

Neo-classical economists such as Jevons, Manager, Walras and Marshall considered that utility functions were additive. However,in real life, this is not so. In actual life, the utility or satisfaction derived from a good depends upon the availability of some other goods which may be either substitute for or complementary to each other. For example, the utility derived from a pen depends upon whether ink is available or not.

On the contrary, if you have only tea, then the utility derived from it would be greater, but if along with tea you also have the coffee, then the utility of tea to you would be comparatively less. Whereas pen and ink are complements of each other, tea and coffee substitute for each other.

It is thus clear that various goods are related to each other in the sense that some are complements of each other and some are substitutes for each other. As a result of this, the utilities derived from various goods are interdependent, that is, they depend upon each other. Therefore, the utility obtained from good is not the function of its quantity alone but also depends upon the existence or consumption of other related goods (complements or substitutes).

It is thus evident that the assumption of the independence of utilities by Marshall and other supporters of marginal utility analysis is a great defect and shortcoming of their analysis. As we shall see below, the hypothesis of independent utilities along with the assumption of constant marginal utility of money reduces the validity of Marshallian demand theorem to the one-commodity model only.

2.11.3 The assumption of constant marginal utility of money is not valid

An important assumption of cardinal utility analysis is that when a consumer spends the varying amount on a good or various goods, or when the price of a good changes, the marginal utility of money remains unchanged. However, in actual practice, this is not correct. As a consumer spends his money income on the goods, money income left with him declines.

With the decline in money income of the consumer as a result of an increase in his expenditure on goods, the marginal utility of money to him rises. Further, when the price of a commodity changes, the real income of the consumer also changes. With this change in real income, the marginal utility of money will change, and this would affect the demand for the good in question, even though the total money income available to the consumer remains the same.

However, utility analysis ignores all this and does not take cognizance of the changes in real income and its effect on demand for goods following the change in the price of a good. As we shall see below, it is because of the assumption of constant marginal utility of money that Marshall ignored the income effect of the price change which prevented Marshall from understanding the composite character of the price effect (that is, price effect is the sum of substitution effect and income effect).

Moreover, as we shall see later, the assumption of constant marginal utility of money together with the hypothesis of independent utilities renders the Marshall's demand theorem to be valid in case of one commodity. Further, it is because of the constant marginal utility of money and therefore the neglect of the income effect by Marshall that he could not explain Giffen Paradox.

According to Marshall, utility from a good can be measured in terms of money (that is, how much money a consumer is prepared to sacrifice for a good). However, to be able to measure utility in terms of money, marginal utility of money itself should remain constant. Therefore, the assumption of constant marginal utility of money is very crucial to Marshallian demand analysis. By constant marginal utility of money, Marshall could assert that "utility is not only measurable in principle" but also "measurable in fact."

However, as we shall see below, in case a consumer has to spread his money income on some goods, there is a necessity for a revision of marginal utility of money with every change in the price of a good. In other words, in a multi-commodity model, marginal utility of money does not remain invariant or constant.

Now, when it is realized that marginal utility of money does not remain constant, then Marshall's belief that utility is 'measurable in fact' in terms of money does not hold good. However, if in marginal utility analysis, the utility is conceived only to be 'measurable in principle' and not in fact, then it practically gives up cardinal measurement of utility and comes near to the ordinal measurement of utility.

2.11.4 *Marshallian demand theorem cannot genuinely be derived except in a one-commodity case*

J.R. Hicks and Tapas Majumdar have criticized Marshallian utility analysis on the ground that "Marshallian demand theorem cannot genuinely be derived from the marginal utility hypothesis except in a one-commodity model without contradicting the assumption of constant marginal utility of money." In other words, Marshall's demand theorem and the constant marginal utility of money are incompatible except in a one-commodity case. As a result, Marshall's demand theorem cannot be validity derived in the case when a consumer spends his money on more than one good.

In order to know the truth of this assertion, consider a consumer who has a given amount of income to spend on some goods with given prices. According to utility analysis, the consumer will be in equilibrium when he is spending money on goods in such a way that the marginal utility of each good is proportional to its price. Let us assume that, in his equilibrium position, the consumer is buying the q_1 quantity of a good *X* at a price P_1. The marginal utility of good *X*, in his equilibrium position, will be equal to its price p_1 multiplied by the marginal utility of money (which, in Marshallian utility analysis, serves as the unit of measurement).

2.11.5 *Thus, in the equilibrium position, the following equation will be fulfilled*

$$MU_x / = MU_m x p_1 \qquad (2.12)$$

Since the consumer is buying the q_1 quantity of good *X* at price P_1, he will be spending P_1Q_1 amount of money on it. Now, suppose that the price of good *X* rises from p_1 to p_2. With this rise in the price of *X*, all other things remaining the same, the consumer will at once find himself in disequilibrium state, for the marginal of good *X* will now be less than the higher price p_g multiplied by the marginal utility of money (Mu_m) which is assumed to remain unchanged and constant. Thus, now there will be

$$MU_x \prec MU_m \bullet p_2 \qquad (2.13)$$

In order to restore his equilibrium, the consumer will buy less of good *X* so that the marginal utility of good *X* (MU_x) would rise and become equal to the product of p_2 and MU_m. Suppose in this new equilibrium position, he is buying q_2 of good *X*, which will be less than q_1. With this, he will now be spending the p_2q_2 amount of money on good *X*. Now, the important thing to see is that whether his new expenditure p_2q_2 on good *X* is equal to, smaller or greater than $P_1 \cdot q_1$.

This depends upon the elasticity of the marginal utility curve, i.e. the price elasticity of demand. If the elasticity of marginal utility curve of good *X* is unity, then the new expenditure on good *X* (i.e.p_2q_2) after the rise in its price from p_1 to p_2 will be equal to the initial expenditure p_1q_1. When the monetary expenditure made on the good remains constant as a result of a change in price, then the Marshallian theory is valid.

However, constant monetary expenditure following a price change is only a rare phenomenon. However, the Marshallian demand theory breaks down when the new expenditure p_2q_2, after the rise in price, instead of being equal, is smaller or greater than the initial expenditure p_2q_2.

If the elasticity of marginal utility curve is greater than one (that is, price demand for the good is elastic), then the new expenditure p_2q_2, after the rise in price from p_1 to p_2, will be less than the initial expenditure *p*. On the other hand, if the elasticity of the marginal utility curve is less than unity, then the new expenditure p_2q_2 after the rise in price will be greater than the initial expenditure p_1q_1.

Now, if the new expenditure $\mathrm{p_2q_2}$ on good *X* is less than the initial expenditure p_1q_1 on it, it means that more money will be left with the consumer to spend on goods other than *X*. And if the new expenditure p_2q_2 on good *X* is greater than the initial expenditure $\mathrm{p_1q_1}$ on it, then less money would be left with him to spend on goods other than *X*.

To ensure that the consumer spends the entire amount of money available to him, then in case of new expenditure p_2q_2 on good *X* being smaller or greater than initial expenditure p_1q_1 on it, the expenditure or goods other than *X* and therefore consumer's demand for them will change.

However, in Marshallian theoretical framework, this further adjustment in consumer's expenditure on goods other than *X* can occur only if the unit of utility measurement, that is, the marginal utility of money, revised or changed. However, Marshall assumes marginal utility of money to remain constant.

Thus, we see that the marginal utility of money cannot be assumed to remain constant when the consumer has to spread his money income on some goods. In case of more than one good, Marshallian demand theorem cannot be genuinely derived while keeping the marginal utility of money constant.

If in Marshallian demand analysis, this difficulty is avoided by giving up the assumption of constant marginal utility of money, then money can no longer provide the measuring rod, and we can no longer express the marginal utility of a commodity in units of money. If we cannot express marginal utility in terms of common numeraire (which money is defined to be), the cardinality of utility would be devoid of any operational significance.

Only in case, there is one good on which the consumer has to spend his money, Marshallian demand theorem can be validity derived. To conclude, in the words of Majumdar,

> Except in a strictly one-commodity world, therefore, the assumption of a constant marginal utility of money would be incompatible with the Marshallian demand theorem.

Without the assumption of an invariant unit of measurement, the assertion of measurability would be entirely meaningless. The necessity and the possibility of a revision of the unit of utility measurement, following every change in price, had been assumed away in Marshallian theory under cover of'other things remaining the same' clause.

2.11.6 Cardinal utility analysis does not split up the price effect into substitution and income effects

The third shortcoming of the cardinal utility analysis is that it does not distinguish between the income effect and the substitutional effect of the price change.

We know that when the price of good falls, the consumer becomes better off than before, that is, a fall in the price of a good brings about an increase in the real income of the consumer. In other words, if with the fall in price the consumer purchases the same quantity of the good as before, then he would be left with some income.

With this income, he would be in a position to purchase more of this good as well as other goods. This is the income effect of the fall in price on the quantity demanded of a good. Besides, when the price of a good falls, it becomes relatively cheaper than other goods, and as a result, the consumer is induced to substitute that good for others. This result increases in quantity demanded of that good. This is the substitution effect of the price change on the quantity demanded of the good.

With the fall in the price of a good, the quantity demanded of it rises because of income effect and substitution effect. However, cardinal utility analysis does not make clear the distinction between the income and the substitution effects of the price change. Marshall and other exponents of marginal utility analysis ignored the income effect of the price change by assuming the constancy of the marginal utility of money. Thus, according to Tapas Majumdar, "the assumption of constant marginal utility of money obscured Marshall's insight into the truly composite character of the unduly simplified price-demand relationship."

They explained the changes in demand as a result of a change in the price of a good by substitution effect on it. Thus, marginal utility analysis does not tell us about how much quantity demanded increases due to income effect and how much due to substitution effect as a result of the fall in price of a good J.R. Hicks rightly remarks, "the cardinal theory accordingly leaves that distinction between income effect and substitution effect of a price change as an empty box which is crying out to be filled." In the same way, Tapas Majumdar says, "The efficiency and precision with which the Hicks-Allen approach can distinguish between the income and sub estimation effects of a price change leaves the cardinal argument in a very poor state indeed."

2.11.7 Marshall could not explain Giffen Paradox

By not visualizing the price effect as a combination of substitution and income effects and ignoring the income effect of the price change, Marshall could not explain the Giffen Paradox. He treated it merely as an exception to his law of demand. In contrast to it, Indifference Curve analysis has been able to explain the Giffen good case satisfactorily.

According to Indifference Curve analysis, in case of a Giffen Paradox or the Giffen, good negative income effect of the price change is more powerful than substitution effect so that when the price of a Giffen good falls, the negative income effect outweighs the substitution effect with the result that quantity demanded of it falls.

Thus, in case of a Giffen good, quantity demanded varies directly with the price and the Marshall's law of demand does not hold good. It is because of the constant marginal utility of money and therefore the neglect of the income effect of price change that Marshall could not explain why the quantity demanded of the Giffen good falls when its price falls and rises when its price rises. This is a serious lacuna in Marshal Elian's utility analysis of demand.

2.11.8 Marginal utility analysis assumes too much and explains too little

Marginal utility analysis is also criticized on the ground that it takes more assumptions and also more severe ones than those of ordinal utility analysis of Indifference Curve technique. Marginal utility analysis assumes, among others, that utility is cardinally measurable and also that marginal utility of money remains constant. Hicks-Allen's Indifference Curve analysis does not take these assumptions, and even then, it is not only able to deduce all the theorems which cardinal utility analysis can but also deduces a more general theorem of demand.

In other words, Indifference Curve analysis explains not only that much as cardinal utility analysis does but even goes further and that too with fewer and less severe assumptions. Taking less severe assumption of ordinal utility and without assuming constant marginal utility of money, the analysiscan arrive at the condition of consumer's equilibrium: namely, equality of marginal rate of substitution (MRS) with the price ratio between the goods, which is similar to the proportionality rule of Marshall. Further, since Indifference Curve analysis does not assume constant marginal utility of money, it can derive a valid demand theorem in a more than one-commodity case.

In other words, Indifference Curve analysis dearly explains why in the case of Giffen, goods quality demanded increases with the rise in price and decreases with the fall in price. Indifference Curve analysis explains even the case of ordinary inferior goods (other than Giffen goods) in a more analytical Inner.

It may be noted that even if the valid demand is derived for the Marshallian hypothesis, it would still be rejected because "better hypothesis" of indifference preference analysis was available which can enunciate more general demand theorem (covering the case of Giffen goods) with fewer, less severe and more realistic assumptions.

Because of the above drawbacks, cardinal utility analysis has been given up in modern economic theory and demand is analyzed with new approaches to demand theory.

2.12 Usefulness of the Ordinal Theory

The utility of the ordinal theory suggests that utility is not as measurable as price and quantity, but it is possible that a person ranked different products in terms of desirability, which means that based on this theory, although it cannot be said that desirability is a commodity, it is possible to say that the utility of A is greater than the utility of B, and vice versa.The point that it seems necessary is that the law of descendants of the final utility applies both to the usefulness of the cardinality and to the utility of the arbitrariness.

To better understand the subject of this theory, consider the following article, provided by economicsconcepts.com

2.12.1 *Theory of Ordinal Utility/Indifference Curve Analysis*

2.12.1.1 *Definition and Explanation*

The ***Indifference Curve*** indicates the various combinations of two goods which yield equal satisfaction to the consumer. By definition,

"An Indifference Curve shows all the various combinations of two goods that give an equal amount of satisfaction to a consumer."

The ***Indifference Curve analysis approach*** was first introduced by Slustsky, a Russian Economist in 1915. Later, it was developed by J.R. Hicks and R.G.D. Allen in the year 1928.

These economists are of the opinion that it is wrong to base the theory of consumption on two assumptions.

- i There is only one commodity which a person will buy at one time.
- ii The utility can be measured.

Their point of view is that ***utility*** is purely subjective and is immeasurable. Moreover, an individual is interested in a combination of related goods and in the purchase of one commodity at one time. So, they base the theory of consumption on the scale of preference and the ordinal ranks or order their preferences.

2.12.1.2 *Assumptions*

The ***ordinal utility theory or the Indifference Curve analysis*** is based on six main assumptions:

- i ***Rational behavior of the consumer.*** It is assumed that individuals are rational in making decisions from their expenditures on consumer goods.
- ii ***The utility is ordinal.*** The utility cannot be measured cardinally. It can be, however, expressed ordinally. In other words, the consumer can rank the basket of goods according to the satisfaction or utility of each basket.
- iii ***The diminishing marginal rate of substitution.*** In the Indifference Curve analysis, the principle of diminishing marginal rate of substitution is assumed.

iv ***Consistency in choice.*** The consumer, it is assumed, is consistent in his behavior during a period. For insistence, if the consumer prefers combinations of A of good to the combinations B of goods, he then remains consistent in his choice. His preference, during another period, does not change. Symbolically, it can be expressedas

$$\text{If} \quad A > B, \quad \text{then} \quad B > A \tag{2.14}$$

v ***Consumer's preference not self-contradictory.*** The consumer's preferences are not self-contradictory. It means that if combination A is preferred over combination B is preferred over C, then combination A is preferred over combination A is preferred over C. Symbolically, it can be expressed

$$\text{If} \quad A > B \quad \text{and} \quad B > C, \quad \text{then} \quad A > C \tag{2.15}$$

vi ***Goods consumed are substitutable.*** The goods consumed by the consumer are substitutable. The utility can be maintained at the same level by consuming more of some goods and less of the other. There are many combinations of the two commodities which are equally preferred by a consumer, and he is indifferent as to which of the two he receives.

2.12.1.3 Example

For example, a person has a limited amount of income which he wishes to spend on two commodities, rice and wheat. Let us suppose that the following commodities are equally valued by him.

Various Combinations

A	16 Kilograms of Rice	Plus	2 Kilograms of Wheat
B	12 Kilograms of Rice	Plus	5 Kilograms of Wheat
C	11 Kilograms of Rice	Plus	7 Kilograms of Wheat
D	10 Kilograms of Rice	Plus	10 Kilograms of Wheat
E	9 Kilograms of Rice	Plus	15 Kilograms of Wheat

It is a matter of indifference for the consumer as to which combination he buys. He may buy 16 kilograms of rice and 2 kilograms of wheat or 9 kilograms of rice and 15 kilograms of wheat. He equally prefers all these combinations.

An Indifference Curve thus is composed of a set of consumption alternatives each of which yields the same total amount of satisfaction. These combinations can also be shown by an Indifference Curve.

2.12.1.4 Figure/Diagram of the Indifference Curve

The consumer's preferences can be shown in a diagram with an Indifference Curve. The indifference was showing nothing about the absolute amounts of satisfaction obtained. It merely indicates a set of consumption bundles that the consumer views as being equally satisfactory.

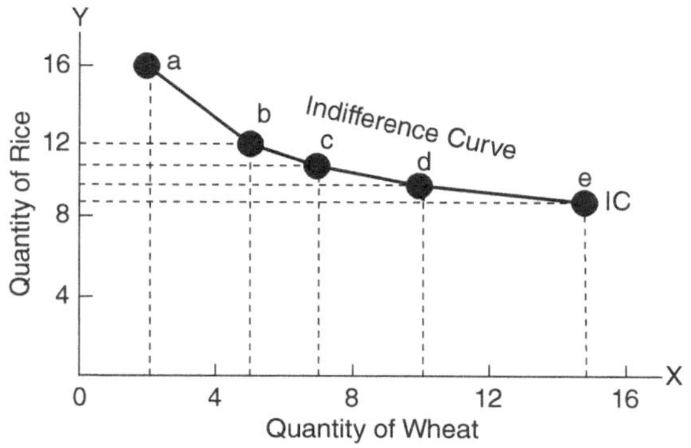

Figure 2.4 Indifference Curve.

In Figure 2.4, we measure the quantity of wheat along the *X*-axis (in kilograms) and along the *Y*-axis, the quantity of rice (in kilograms). IC is an Indifference Curve.

It is shown in the diagram that a consumer may buy 12 kilograms of rice and 5 kilograms of wheat or 9 kilograms of rice and 15 kilograms of wheat. Both these combinations are equally preferred by him,and he is indifferent to these two combinations. When the scale of preference of the consumer is graphed, by joining the points a, b, c, d and e, we obtain an Indifference Curve IC.

Every point on the Indifference Curve represents a different combination of the two goods, and the consumer is indifferent between any two points on the Indifference Curve. All the combinations are equally desirable to the consumer. The consumer is indifferent as to which combination he receives. The Indifference Curve IC thus is a locus of different combinations of two goods which yield the same level of satisfaction.

2.12.1.5 An Indifference Map

A graph showing a whole set of Indifference Curves is called an ***indifference map.*** An indifference map, in other words, is comprised of a set of Indifference Curves. Each successive curve further from the original curve indicates a higher level of total satisfaction.

In Figure 2.5, three Indifference Curves IC^1, IC^2 and IC^3 have been shown. The various combinations of goods of wheat and rice lying on IC^1 yield the same level of satisfaction to the consumer. The combinations of goods lying on higher Indifference Curve IC^2 contain more both the goods, wheat and rice. The Indifference Curve IC^2 gives more satisfaction to the consumer than IC^1. Similarly, the set of combinations of two goods on IC^3 yields still higher satisfaction to the consumer than IC^2. In short, the further away a particular curve is from the origin, the higher level of satisfaction it represents.

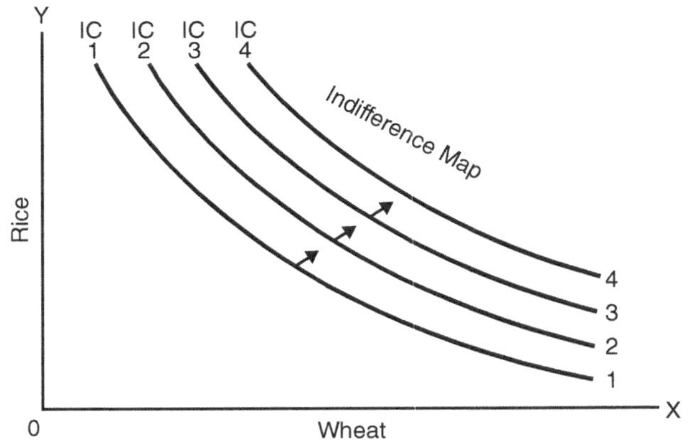

Figure 2.5 Indifference Map.

It may be noted here that while an Indifference Crve shows all those combinations of wheat and rice which provide equal satisfaction to the consumer, it does not indicate exactly how much satisfaction is derived by the consumer from these combinations. It is because the concept of ordinal utility does not involve the qualitative measurement of utility.

2.13 Total Desirability and Final Utility

A person makes a specific item because he or she wants to use it for satisfaction or desirability. The larger the unit of a commodity consumed by a person in a time unit, the greater the overall utility he earns. As a result, the overall utility increases, but the additional utility or final utility obtained by consuming each additional unit of the product usually decreases. At some levels of consumption, the optimal use of the individual as a result of the consumption of the product will be maximized, and at this level, his final utility will be zero. This level is called the saturation point. The consumption of additional units from the goods reduces the overall utility and negativity of the final utility. According to the above, the utility function can be plotted as follows.

If we consider two products x and y, then the utility function is in the form $U(x, y)$. Figure 2.6 shows the utility function $U(x, y)$ in a three-dimensional graph. In this graph, the values of x, $y$are measured on the axes that are located on the horizontal plane; the utility is also measured in terms of utile and on the perpendicular axis. It should be noted that we are here with a level of utility, and the higher the amount of goods, the higher this level also rises.

Figure 2.7 shows a different representation of the utility function that is discussed above. The curves drawn here at the level of desirability of the total are obtained by connecting the points of the tent of utility in Figure 2.8 with equal height. So, all the points of each of these lines have the same and constant utility.

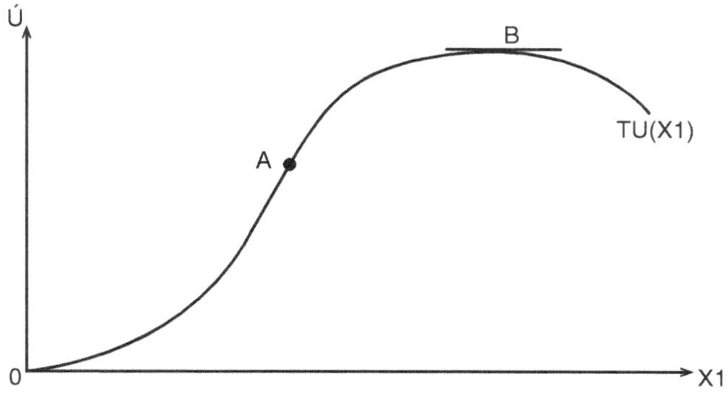

Figure 2.6 Utility function.

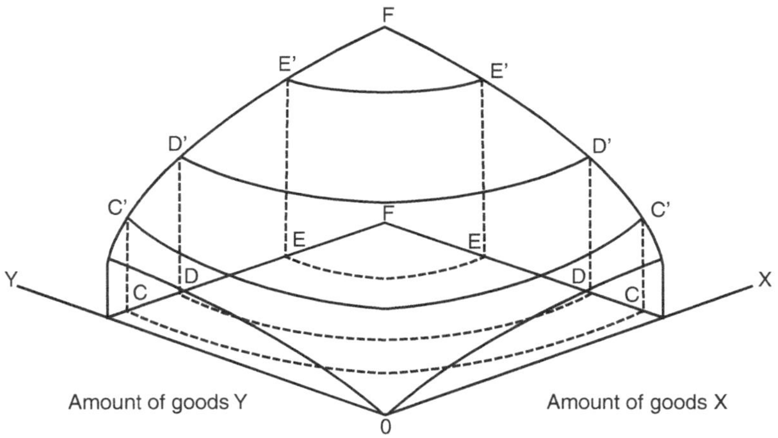

Figure 2.7 Utility function for two goods.

Now, if we look at the lines (i.e. Ć, ĎĎ, ĖĖ) from above, we can see curves like those in Figure 2.8. In other words, if we plot these lines on the screen, curves such as EE, DD and CC are obtained such that the utility value is fixed on each of them.

To each of these curves, the curvature of indifference and their sum is called an indifference map. Therefore, each of these Indifference Curves will show the different combinations of two products *x*, *y*, which offer the same satisfaction to the consumer. As we see in the figure, we find that the Indifference Curves from left to right have a negative slope, do not cut each other and are convex to the origin of the coordinates, while the higher Indifference Curves indicate a higher utility than the curves in which the difference is lower than themselves (Figure 2.9).

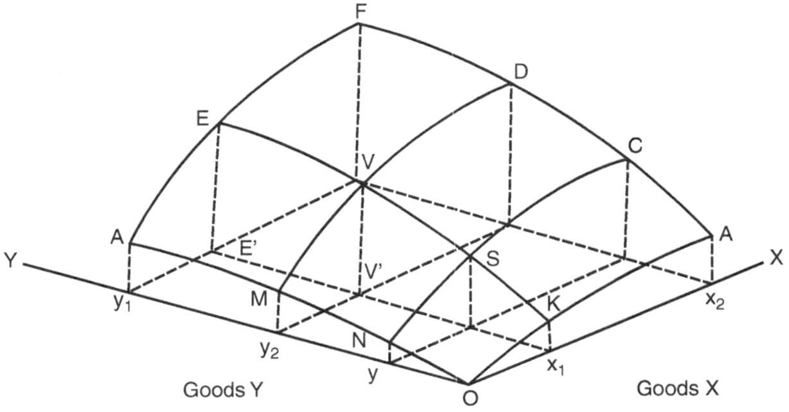

Figure 2.8 The utility function of the two commodities and the extraction of Indifference Curves.

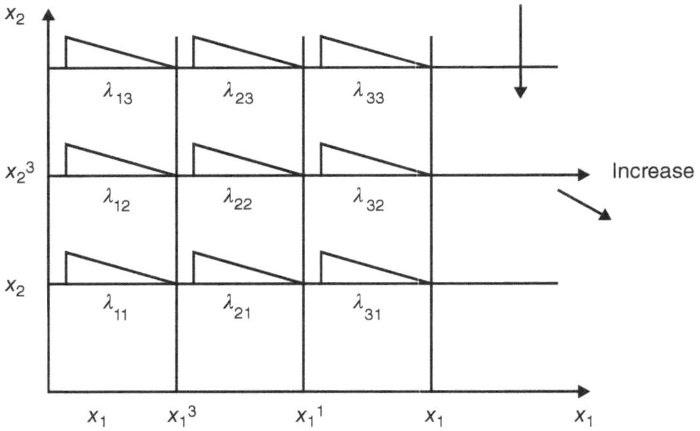

Figure 2.9 Relationship of the final rate of succession of two commodities.

2.14 The final rate of succession

The marginal rate of x succession for y (MRS_{xy}) represents the amount of y that a consumer wishes to lose in order to obtain a unit of x while remaining on the curvature of indifference.

As the person moves down the Indifference Curve, the MRS_{xy} decreases, resulting in a coordinate axis.

$$
\begin{array}{ll}
\lambda_{31} > \lambda_{21} > \lambda_{11} & \lambda_{11} > \lambda_{12} > \lambda_{13} \\
\lambda_{32} > \lambda_{22} > \lambda_{12} & \lambda_{21} > \lambda_{22} > \lambda_{23} \\
\lambda_{33} > \lambda_{23} > \lambda_{13} & \lambda_{31} > \lambda_{32} > \lambda_{33}
\end{array}
\tag{2.16}
$$

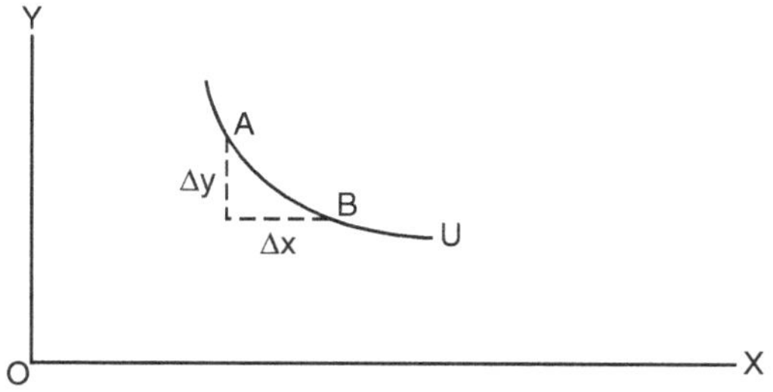

Figure 2.10 Indifference Curve.

2.15 The Relationship between Final Rate of Substitution with Final Utility

Another important relationship to be considered is that the absolute value of the slope of the Indifference Curve is equal to the final utility *x* ratio to the final utility of *y*. (Commodity *x* is located on the horizontal axis.) To prove, suppose the final utility of *x* is equal to Mux, and the ultimate utility of *y* is Muy. If the consumer moves from point A to point B in the form of apathetic u in Figure 2.10, the consumption of commodity *x* increases by Δx, and the consumption of commodity *y* decreases by Δy. As a result of an increase in the consumption of *x*, the utility of the consumer increases by an amount of $\text{Mux}.\Delta x$, while as a result of decreasing consumption, the amount of utility of the product decreases to $\text{Muy}.\Delta y$.

If we consider that by moving the Indifference Curve u, the amount of consumer utility is constant, then the net change in utility should be zero, so we can write,

$$\begin{aligned} &\Delta y.\text{Muy} + \Delta x.\text{Mux} = 0 \\ &|\Delta y/\Delta x| = \text{MU}_x/\text{MU}_y \end{aligned} \tag{2.17}$$

However, for small values Δy and Δx, $|\Delta y/\Delta x|$ is the same as the absolute slope of the curvature of indifference. So, the following important result is obtained:

$$\text{MRS}_{xy} = |\Delta y / \Delta x| = \text{MU}_x/\text{MU}_y \tag{2.18}$$

Who has researched the theory of utility?

Kinney and Wiruth (1976) studied the utility theory and its rules. In their theory, all essential and functional aspects are considered, the accuracy of the data and the possibility of risk aversion. In their design, the evaluation is based on the multi-dimensional utility theory. This theory allows the designer to examine the priorities in the specified dimensions and achieve a single design (a plan that results in individual opinions). Utility theory helps the decision maker to be light and heavy among variables, and then reaches a single result.

Hershey and Schoemaker (1985) considers utility theory to be more than the odds of choosing between some variables. In his view, when our information is complete about a problem, the probability of error is reduced, and more information can be chosen to target the target. However, despite incomplete information, the probability of error in decision making between variables increases.

The hierarchical analysis process (AHP) is a common decision making method in complex and multi-dimensional issues. In this way, the quantitative and qualitative options of the problem are examined simultaneously. Moore (1998a,b) questioned the authenticity of this method and suggested reasons for its illogicality.

In contrast, according to Callaghan and Lewis, in the theory of utility and AHP, which are referred to as value methods (the ways in which the decision-maker among a number of variables achieves a general result) since utility applications are not constant, the interpersonal relationships of the group can be effective in individual decisions and thus in the final decision of the group.

James-El-Rogers has presented his subject for a passenger plane design.

(https://www.google.com/url?sa=t&source=web&rct=j&url=http://aircraftdesign.nuaa.edu.cn/MDO/ref/overview/AIAA_FD99.slides.pdf&ved=2ahUKEwiUn6Wpk87kAhWC4J4KHfXQCXwQFjAAegQIBhAB&usg=AOvVaw1GICeZEa2iIFdMOY_7eXwl). The summary of his plan is included below.

In designing a passenger aircraft design, the challenge is subject to the following constraints:

- Low cost
- Lowest weight
- The largest passenger capacity
- Maximum extent
- Highest speed

So that.

$$\begin{aligned}&\text{Overall cost} = \text{Engine cost (engine/wing) 2)} +\\&\text{Aircraft cost} + \text{(Wing cost) 2}\\&\text{Overall weight} = \text{Total engine weight} + \text{Airplane body weight} +\\&\text{Overall weight of the wings} + \text{Fuel weight Speed 738}\end{aligned} \tag{2.19}$$

The aircraft is divided into three sub-devices.

Engine, body and wing. Each subsystem has four attributes as shown in Table 2.4.

There are five options for each of these subsystems whose properties are specified in Tables 2.5–2.7.

Table 2.4 Subsystem

The wings	*The body*	*Engine*
The weight of each wing	Body weight	The weight of each engine
The cost of building each wing	Body cost	Cost per engine
Fuel consumed each wing	Capacity	Speed
The necessary engine for each wing	Breadth	Fuel consumption

Table 2.5 Existing engine specifications

Engine	*1*	*2*	*3*	*4*	*5*
WeightIbs	11,000	12,000	4,000	13,000	5,000
Cost/million dollars	5	4.5	1.5	3.5	1
Speed	0.4	0.3	0.2	0.35	0.1
Consumption	200	100	50	150	20

Table 2.6 Existing body specifications

The body	*1*	*2*	*3*	*4*	*5*
Weight	55,000	36,000	20,000	25,000	11,000
Cost/million dollars	6	4.5	2.5	4	1.5
Capacity	400	300	350	200	100
Breadth	400	2,600	1,200	3,000	1,000

Table 2.7 Existing wing specifications

Wing	*1*	*2*	*3*	*4*	*5*
Weight	25,000	20,000	15,000	10,000	5,000
Cost/million dollars	2.5	2	1.5	1	0.5
Fuel consumption	2,500	2,000	1,500	1,000	500
Engine for each wing	2	1	1	2	2

In this study, an aircraft is assessed based on five criteria: 1– Cost, 2– Weight, 3– Speed, 4– Size, and 5 – Number of passengers (Figure 2.11).

Questions are asked in the process of evaluating the middle categories of the decision maker regarding the extent of the changes that are more important to him. These questions can be as follows:

Which change is more important?

A 4.5 million to 23.55 million
B 23.75 million to 43 million

Assuming that the decision maker chooses change B, the next question is whether between (a) 4.5 million to 30 million and change (b) between 30 million and 43 million is preferred? If at this stage, the decision maker finds that there is no difference between changes A and B, the midpoint of 23.53 million is determined, and 0.5 is the optimal point. This is the desired point, and 43 million combined to function as a desirable measurement (Figure 2.12).

These questions are also asked of other decision makers, and the utility of each one is extracted; then, the combination of these functions is obtained in the form of multiple utility functions. Finally, using these utility functions, the

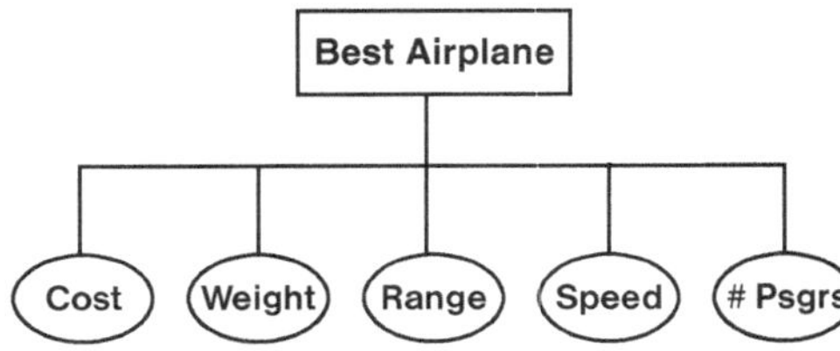

Figure 2.11 The criteria for choosing the best aircraft.

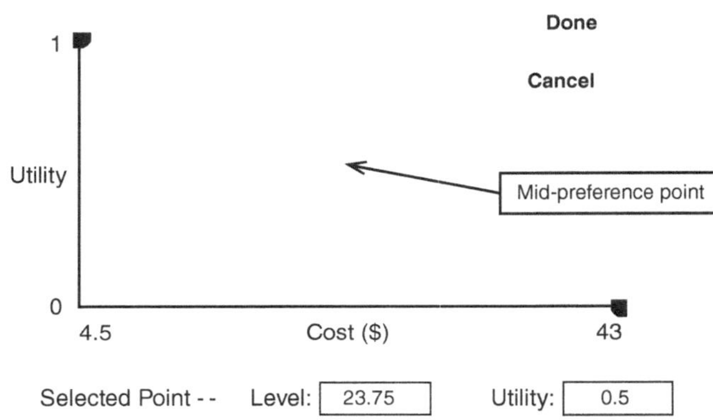

Figure 2.12 Preference point.

relative weights of the criteria are extracted, and ultimately the final weight of the options is determined.

In the end, the general manager chooses the final option of buying each item, emphasizing the dictatorial theory of goodwill.

3 Models of Risks in Economics

3.1 Using Analyzing Hierarchical Process

The hierarchical analysis process, as mentioned in the previous discussions, has several steps that each task is responsible for as a whole. The process of this process is such that each step is considered as the next step. The steps are in series. These steps are:

A *Step one.* The Hierarchy
B *Step two.* Calculate the Weight
C *Step three.* System Compatibility

In this process, the health and safety of each stage are necessary, and the desired outcome is acceptable to the next stage. Therefore, inaccuracy in each step causes an incorrect process until the end of the process along with the data to be analyzed, and in this case, the results will be scientifically rejected.

In this book, we tried to make our relative weights between the criteria and the sub-criteria instead of our linear relationship between them and the non-linear relations, especially the utility function relations, for example, when the choice of automobile is the goal of decision making (DM).

To compare the criteria in a paired comparison matrix, a constant number is used.

$$\begin{bmatrix} 1 & \frac{1}{4} & \frac{1}{4} \\ 4 & 1 & \frac{1}{4} \\ 3 & 3 & 1 \end{bmatrix} \tag{3.1}$$

However, an important point in the matrix above is as follows.

As the convenience of a car is increased, the decision maker's utility becomes less convenient, for example, if the unit is a convenient unit of the person x, if a convenience unit is added, the comfort convenience of the decision maker is smaller than x, the clearer it is if it is supposed to be. For each unit of convenience, the decision maker does not take a certain amount of safety. Certainly, for the second unit, it is unlikely to lose the safety of the previous time.

The main purpose of the author in this book is that, since the desirability of any criterion in the event of its increase or decrease, in order to calculate our relative weight between the criteria and sub-criteria, instead of placing a fixed number in the pairwise comparison matrix, the utility function will be used as the result of the algorithm of the hierarchical analysis process as follows.

A hierarchical problem is assumed to be the following (Figure 3.1):

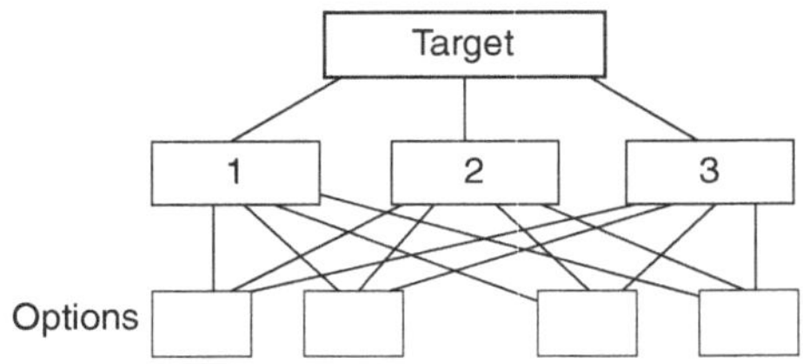

Figure 3.1 A given hierarchy.

Concerning the function relations for criteria 1 and 2, we have Figure 3.2.

$$\begin{array}{ll} \lambda_{31} > \lambda_{21} > \lambda_{11} & \lambda_{11} > \lambda_{12} > \lambda_{13} \\ \lambda_{32} > \lambda_{22} > \lambda_{12} & \lambda_{21} > \lambda_{22} > \lambda_{23} \\ \lambda_{33} > \lambda_{23} > \lambda_{13} & \lambda_{31} > \lambda_{32} > \lambda_{33} \end{array} \tag{3.2}$$

After extracting the λ values according to the amount λ, x_2, x_1 at different levels using nonlinear function regression, $f(x_1, x_2)$, it is possible to accomplish these

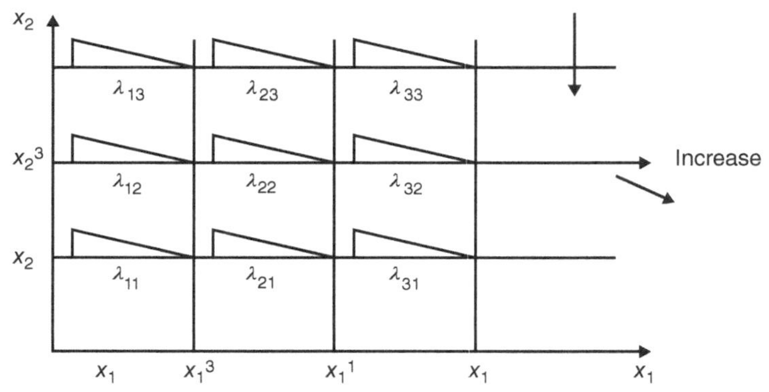

Figure 3.2 Criteria 1 and 2.

steps in the same way as us $f(x_3,x_2), f(x_3,x_1) f(x_2,x_3), f(x_2,x_1), f(x_1,x_3)$. As a result, we have

$$\begin{array}{c} \\ x_1 \\ x_2 \\ x_3 \end{array}\begin{array}{c} \begin{array}{ccc} x_1 & x_2 & x_1 \end{array} \\ \begin{bmatrix} \frac{w^1}{w^1} & \frac{w^1}{w^2} & \frac{w^1}{w^3} \\ \frac{w^2}{w^1} & \frac{w^2}{w^2} & \frac{w^2}{w^3} \\ \frac{w^3}{w^1} & \frac{w^3}{w^2} & \frac{w^3}{w^3} \end{bmatrix} \end{array} \Rightarrow \begin{array}{c} \\ x_1 \\ x_2 \\ x_3 \end{array}\begin{array}{c} \begin{array}{ccc} x_1 & x_2 & x_3 \end{array} \\ \begin{bmatrix} 1 & f(x_1,x_2) & f(x_1,x_3) \\ f(x_2,x_1) & 1 & f(x_2,x_3) \\ f(x_3,x_1) & f(x_3,x_2) & 1 \end{bmatrix} \end{array} \tag{3.3}$$

$$\underset{x_2 \quad x_1}{\text{MRS}} = f(x_1,x_2)$$

In order to calculate and weigh the relative weight of the criteria, we use the least squares method.

$$\min z = \sum_{i=1}^{3}\sum_{j=1}^{3}\left[f\left(x_i,x_j\right)wf - wi\right]^2$$
$$st\sum_{i=1}^{3} wi = 1 \tag{3.4}$$

After the device is solved, the relative weights of the criteria are obtained, which is more accurate than the previous methods. Therefore, with the above, the methodology of the book is as follows:

1 definition of criteria;
2 define and determine the marginal rate of succession (MRS) of each criterion by comparing them to each other;
3 definition of a function for both criteria using the cover curve;
4 determination of Relative Weight of Criteria Using Nonlinear Programming.

Example. The following decision hierarchy is assumed to be the choice of a home (Figure 3.3):

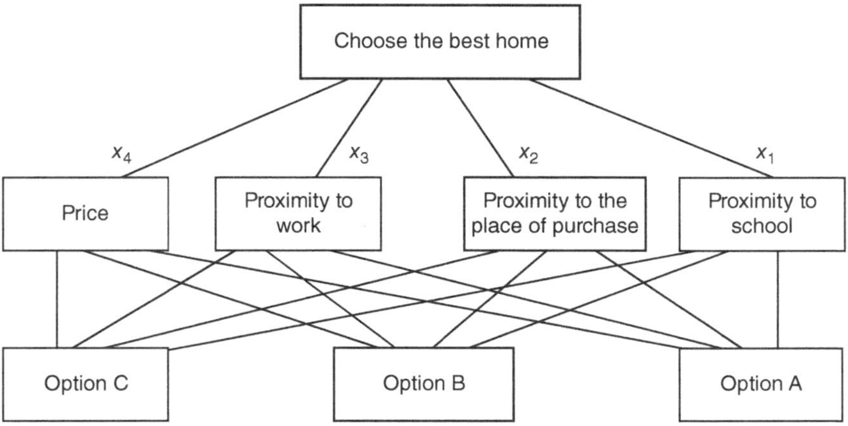

Figure 3.3 A hypothetical problem.

First, based on the analysis process of the hierarchical analysis process, the options are compared in a pairwise manner based on each criterion (Tables 3.1–3.4).

Table 3.1 Analysis result (A)

Proximity to school	*A*	*B*	*C*
A	1	2	8
B	$\frac{1}{2}$	1	6
C	$\frac{1}{8}$	$\frac{1}{6}$	1

Table 3.2 Analysis result (B)

Proximity to the place of purchase	*A*	*B*	*C*
A	1	$\frac{1}{3}$	$\frac{1}{6}$
B	3	1	$\frac{1}{2}$
C	4	2	1

Table 3.3 Analysis result (D)

Price	*A*	*B*	*C*
A	1	4	4
B	3	1	7
C	$\frac{1}{4}$	$\frac{1}{7}$	1

Table 3.4 Analysis result (E)

Proximity to work	*A*	*B*	*C*
A	1	$\frac{1}{4}$	$\frac{1}{2}$
B	4	1	$\frac{1}{3}$
C	6	3	1

Now, the relative weight of each option is obtained based on each criterion, which is used to approximate these weights using the approximate normalization method (Tables 3.5–3.12).

Table 3.5 Proximity to school (A)

Proximity to school	*A*	*B*	*C*
A	1	2	8
B	$\frac{1}{2}$	1	6
C	$\frac{1}{8}$	$\frac{1}{6}$	1
	$\frac{13}{8}$	$\frac{19}{6}$	15

Table 3.6 Proximity to school (B)

Proximity to school	*A*	*B*	*C*	*Average row*
A	0/615	0/631	0/533	0/593
B	0/308	0/316	0/4	0/341
C	0/077	0/053	0/067	0/66
Total sum	1	1	1	1

$$\Rightarrow \text{Proximity to school} \begin{bmatrix} 0/593 \\ 0/341 \\ 0/066 \end{bmatrix}$$

Table 3.7 Proximity to the place of purchase (A)

Proximity to the place of purchase	*A*	*B*	*C*
A	1	$\frac{1}{3}$	$\frac{1}{4}$
B	3	1	$\frac{1}{2}$
C	4	2	1
	8	$\frac{10}{3}$	$\frac{7}{4}$

Table 3.8 Proximity to the place of purchase (B)

Proximity to the place of purchase	*A*	*B*	*C*	*Average row*
A	0/125	0/1	0/143	0/123
B	0/375	0/3	0/286	0/32
C	0/5	0/6	0/571	0/557
Total sum	1	1	1	1

$$\Rightarrow \text{Proximity to the place of purchase}\begin{bmatrix} 0/123 \\ 0/32 \\ 0/557 \end{bmatrix}$$

Table 3.9 Price (A)

Price	*A*	*B*	*C*
A	1	4	4
B	3	1	7
C	$\frac{1}{4}$	$\frac{1}{7}$	1
	$\frac{17}{4}$	$\frac{36}{7}$	12

Table 3.10 Price (B)

Price	*A*	*B*	*C*	*Average row*
A	0/235	0/778	0/333	0/449
B	0/706	0/194	583/0	0/494
C	0/059	0/28	84/0	0/57
Total sum	1	1	1	1

$$\Rightarrow \text{Price}\begin{bmatrix} 0/449 \\ 0/494 \\ 0/57 \end{bmatrix}$$

Table 3.11 Proximity to work (A)

Proximity to work	*A*	*B*	*C*
A	1	$\frac{1}{4}$	$\frac{1}{6}$
B	4	1	$\frac{1}{3}$
C	6	3	1
	11	$\frac{17}{4}$	$\frac{9}{6}$

Table 3.12 Proximity to work (B)

Proximity to work	*A*	*B*	*C*	*Average row*
A	0/09	0/06	0/111	0/087
B	0/36	0/74	0/222	0/274
C	0/55	0/7	0/667	0/639
Total sum	1	1	1	1

$$\Rightarrow \text{Proximity to work} \begin{bmatrix} 0/087 \\ 0/274 \\ 0/639 \end{bmatrix}$$

Now, after gaining the relative weight of the options relative to each criterion, the relative weights of the criteria are compared in a pairwise manner, which is applied in the following way to the use of the utility theory instead of linear relations.

A First, the final rate of substitution of both criteria is obtained in pairs through DM.

So, to extract our utility functions, we have two criteria (Figure 3.4).

$$\begin{array}{l} \text{MRS}: f(x_1, x_2) \\ x_2 \quad x_1 \end{array} \tag{3.5}$$

It is necessary to explain that the values of λ with the question of DM and his placing in different situations are obtained, for example $f(x_1, x_2)$, for such a question. For example if the distance between the school and the workplace be km 25 km10 per kilometer near work is how much a person is willing to be away from school and so on concerningamounts λ, x_2, x_1. The 3D chart and its function can be extracted using the STATISTICA software. Therefore, we have $f(x_1, x_2)$ (Table 3.13).

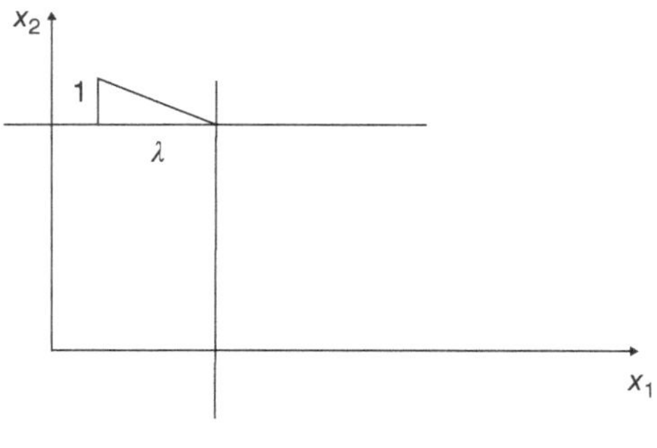

Figure 3.4 The final rate of succession for two pessimistic criteria.

Table 3.13 Extracted function using the STATISTICA software (A)

x_2			
5 km	1	2	3
10 km	2	4	5
15 km	4	6	7

$$f_{(x_1,x_2)}=3/222+0/317x_1-0/2x_2-0/003x_1^2-0/003x_1x_2+0/002x_2^2 \quad (3.6)$$

The same way, $f(x_1,x_3)$, we have Table 3.14.

Table 3.14 Extracted function using the STATISTICA software (B)

x_3			
2 km	2	3	4
3 km	3	5/5	5
5 km	4	6	8
	25 km	15 km	5 km x_1

$$f(x_1,x_3)=2/077+0/136/x_1+3/286x_3-0/005x_1^2 -0/036x_1x_3-0/25x_3^2 \quad (3.7)$$

As a result, we have Table 3.15.

Table 3.15 Extracted function using the STATISTICA software (C)

x_4			
3 km	1	2	3
5 km	2	3	5
6 km	3	6	7
	25 km	15 km	5 km x_1

$$f(x_1,x_4)=6/403+0/05x_1-2/407x_4-0/002x_1^2 -0/032x_1x_4+0/444x_4^2 \quad (3.8)$$

As a result, we have Tables 3.16–3.24.

Table 3.16 Extracted function using the STATISTICA software (D)

x_1			
5 km	1	2	3
15 km	2	4	6
25 km	3	5	7
	15 km	10 km	5 km x_2

$$f(x_2, x_1) = 2/458 - 0/183x_2 + 0/4x_1 - 16x_2^2 - 0/01x_2x_1 - 0/005x_1^2 \tag{3.9}$$

Table 3.17 Extracted function using the STATISTICA software (E)

x_3			
2	$\frac{1}{4}$	$\frac{1}{2}$	1
3	$\frac{3}{2}$	2	3
5	2	4	5

$$f(x_2, x_3) = 5/139 - 0/108x_2 + 3/722x_3 - 0/002x_2^2 - 0/075x_2x_3 - 0/278x_3^2 \tag{3.10}$$

Table 3.18 Extracted function using the STATISTICA software (F)

x_4			
3	10	15	20
5	15	20	25
6	20	25	3
15 km	10 km	5 km	x_2

$$f(x_2, x_4) = 30 - x_2 - 4/167x_4 + 9/869e - 17x_2^2 - 16x_2x_4 + 0/833x_4^2 \tag{3.11}$$

Table 3.19 Extracted function using the STATISTICA software (G)

x_1			
5	5	9	10
15	7	11	15
25	15	19	22
5 km	3 km	2 km	x_3

$$f(x_3, x_1) = 16/599 - 3/349x_3 - 0/071x_1 + 0/222x_3^2 - 0/029x_3x_1 + 0/023x_1^2 \tag{3.12}$$

Table 3.20 Extracted function using the STATISTICA software (H)

x_2			
5	6	14	20
10	13	18	27
15	20	29	33
	5 km	3 km	2 km x_3

$$f(x_3, x_2) = 34/714 - 10/992x_3 + 0/529x_2 + 0/889x_3^2 + 0/021x_3x_2 + 0/04x_2^2 \tag{3.13}$$

Table 3.21 Extracted function using the STATISTICA software (I)

x_4			
3	1	3	5
5	2	4	6
6	3	7	8
	5 km	3 km	2 km x_3

$$f(x_3, x_4) = 13/095 - 1/651x_3 - 3/092x_4 + 0/11x_3^2 - 0/122x_3x_4 + 0/5x_4^2 \tag{3.14}$$

Table 3.22 Extracted function using the STATISTICA software (J)

x_1			
5	2	3	4
15	2/5	4	6
25	3	5	7
	6 km	5 km	3 km x_4

$$f(x_4, x_1) = 0/819 + 1/427x_4 + 0/3x_1 - 0/222x_4^2 - 0/032x_4x_1 - 0/002x_1^2 \tag{3.15}$$

Table 3.23 Extracted function using the STATISTICA software (K)

x_2			
5	1	2	4
10	3	6	8
15	5	8	10
	6 km	5 km	3 km x_4

$$f(x_4, x_2) = 6/556 + 3/127x_4 + 1/333x_2 - 0/444x_4^2 - 0/57x_4x_2 - 0/027x_2^2 \tag{3.16}$$

Table 3.24 Extracted function using the STATISTICA software (L)

x_3			
2	3	6	9
3	7	12	15
5	10	16	19
	6 km	5 km	3 km x_4

$$f(x_4, x_3) = 23/571 + 7/829x_4 + 12/405x_3 - 1/056x_4^2 - 0256x_4x_3 - 1/167x_3^2 \tag{3.17}$$

Now, we have obtained the utility functions (Table 3.25).

Table 3.25 Utility functions

x_1	1	$f(x_1,x_2)$	$f(x_1,x_3)$	$f(x_1,x_4)$
x_2	$f(x_2,x_1)$	1	$f(x_2,x_3)$	$f(x_2,x_4)$
x_3	$f(x_3,x_1)$	$f(x_3,x_2)$	1	$f(x_3,x_4)$
x_4	$f(x_4,x_1)$	$f(x_4,x_2)$	$f(x_4,x_3)$	1

We use the least squares method.

$$\sum_{j=1}^{4}\sum_{j=1}^{4}\left[f(xi, xj)\,wj - wi\right]^2 \quad S.to: \quad \sum wi = 1 \tag{3.18}$$

As a result,

$$\begin{aligned}
&\min\left(\left[3/222+0/317x_1-0/2x_2-0/003x_1^2-0/003x_1x_2+0/002x_2^2\right]w_2-w_1\right)^2\\
&+\left(\left[2/077+0/136x_1-3/28x_2-0/005x_1^2-0/036x_1x_3+0/25x_3^2\right]w_3-w_1\right)^2\\
&+\left(\left[6/403+0/05x_1-2/407x_4-0/002x_1^2-0/032x_1x_4+0/444x_4^2\right]w_4-w_1\right)^2\\
&+\left(\left[2/458+0/183x_2-0/4x_1-16x_2^2-0/01x_2x_1+0/005x_1^2\right]w_1-w_2\right)^2\\
&+\left(\left[5/139+0/108x_2-3/722x_3-0/002x_2^2-0/075x_2x_3+0/278x_3^2\right]w_3-w_2\right)^2\\
&+\left(\left[30-x_2-4/167x_4-17x_2^2-16x_2x_4-0/833x_4^2\right]w_4-w_2\right)^2\\
&+\left(\left[16/599-3/349x_3-0/071x_1+0/222x_3^2-0/023x_3x_1+0/04x_2^2\right]w_1-w_3\right)^2\\
&+\left(\left[34/714-10/992x_3+0/529x_2+0/889x_3^2+0/021x_3x_2+0/04x_2^2\right]w_2-w_2\right)^2\\
&+\left(\left[13/095-1/651x_3-3/092x_4+0/111x_3^2-0/1222x_3x_4+0/5x_4^2\right]w_4-w_3\right)^2\\
&+\left(\left[0/819+1/427x_4+0/3x_1-0/222x_4^2-0/032x_4x_1-0/002x_1^2\right]w_1-w_4\right)^2\\
&+\left(\left[6/556+3/127x_4+1/333x_2-0/444x_4^2-0/057x_4x_2-0/027x_2^2\right]w_2-w_4\right)^2\\
&+\left(\left[23/571+7/829x_4+12/405x_3-1/056x_4^2-0/265x_4x_3-1/167x_3^2\right]w_3-w_4\right)^2
\end{aligned} \tag{3.19}$$

$$\begin{aligned}
st: w_1+w_2+w_3+w_4&=1\\
x_1,x_2,x_3,x_4&\geq 0\\
x_1&\leq 25\\
x_2&\leq 15\\
x_3&\leq 5\\
x_4&\leq 6
\end{aligned}$$

After solving the equation, the relative weight of the criteria is calculated as follows:

$$\begin{aligned}
w_1&=0/56\\
w_2&=0/39\\
w_3&=0/01\\
w_4&=0/04
\end{aligned} \tag{3.20}$$

As a result, we have options for the final weight (Table 3.26).

Table 3.26 Final weight

The relative weight of options	*Proximity to school*	*Proximity to the place of purchase*	*Proximity to work*	*Price*
A	0.593	0.123	0.087	0.449
B	0.341	0.32	0.274	0.494
C	0.066	0.557	0.639	0.57

$$\begin{aligned}&(0/56*0/593)+(0/39*0/123)+(0/01*0/087)+(0/04*0/449)=0/398\\&(0/56*0/341)+(0/39*0/32)+(0/01*0/274)+(0/04*0/494)=0/338\\&(0/56*0/066)+(0/39*0/557)+(0/01*0/639)+(0/04*0/57)=0/283\end{aligned} \tag{3.21}$$

As a result,

$$\begin{aligned}&\text{Option } A = 0/398\\&\text{Option } B = 0/338\\&\text{Option } C = 0/283\end{aligned} \tag{3.22}$$

Considering the results of the proposed model's performance and considering the individual components discussed, this method and the utility theory in order to obtain our relative weights can be declared between the criteria and the sub-criteria.

In the new method, in comparison with the previous method, the accuracy and reliability of the final option can be increased in the hierarchical analysis process. Therefore, the proposed method is used to solve the hierarchical analysis process problems.

In this book, the hierarchy analysis process has been tried and evaluated, with an emphasis on strengths and providing solutions to the weaknesses of a new, more reliable method. Therefore, the new model is summarized as follows:

1 definition of criteria;
2 define and determine the MRS of each criterion by comparing them to each other;
3 definition of a function for both criteria using the cover curve;
4 determination of Relative Weight of Criteria Using Nonlinear Programming.

The main purpose of this book is to improve the process of analyzing hierarchical AHP (analytic hierarchy process) using our nonlinear relationship between criteria and sub-criteriabecause the author believes that this research work is not free of any errors or mistakes. What is presented is the result of his research during the writing of this book and recommends the following suggestions.

1 Applying the utility theory to other DM techniques in which the criteria of the problem under discussion are compared, especially in techniques in which an increase in one criterion leads to a reduction in the other criterion.
2 The method of extracting data from DM in the form of a questionnaire or other tools to obtain the marginal rate of substitution of both criteria can be considered as a research work.
3 Study and research on the application of nonlinear relationships to obtain our relative weights between criteria and sub-criteria can also be considered as a research work.
4 Given that the proposed method of the DM matrix is not proposed for calculating the incompatibility rate, providing a method for calculating the inconsistency rate in a case where the comparison of criteria and sub-criteria is nonlinear can be considered as an investigative work. To be placed.

What is the decision?

If we consider different management activities, it is seen that the essence of the activities of decision-making management is. DM is an integral part of management, and it is a manifestation in every management task. In defining the organization's policies, in defining goals, designing an organization, selecting, evaluating, and in all verbs and actions of decision-making management, is a key component. The manager is always faced with issues that he or she is asking for a decision, and the quality and manner of these decisions determine the extent to which the goals of the organization are met. Hence, it is important to understand the methods of DM and the knowledge of DM techniques for managers, and to use these methods and tools to increase the ability of managers to make more efficient and effective decisions.

3.2 Decision Making (DM) process

Herbert Simon, one of the management scholars, considers DM as the essence of management. In his opinion, management is nothing but a decision, and a successful manager can be a successful decision maker.

Chester Barnard, one of the decision makers, says, "Decision making is the core of the manager's duties, and the manager's skill in decision making is apparent in the performance of his duties and in the quality of the services he offers."

In a very simple definition, DM is the choice of a way through a variety of ways. As it is understood, the main task of the decision maker is to get the possible ways and the results from them, and choose the best of them, and if he can make this choice in the right way, his decisions will be effective and constructive. The decision maker may make decisions by resorting to the powers of metaphysics, experience, illumination, or accident, but the main purpose here is to give a brief overview of the methods and techniques of DM that allows the manager to make practical decisions. Moreover, he helps him get quick and accurate decisions.

In general, the DM process can include the following six steps.

The first step involves identifying the problem that has created a barrier to achieving the goal. At this stage, the original and actual problem must be tried and understood to be properly defined.

The next step is to find possible solutions to resolve the issue. The list of solutions is based on the scientific and practical experience of the manager and the information and statistics available to him. The more solutions are found to solve the problem, the better the decision will be made.

The third step is to select a benchmark for measuring and evaluating possible solutions. In order for solutions to be judged against each other, we need to examine them by the standard. For example, you can consider the cost or profit as a benchmark for possible assessments, and then measure which solutions will cost less or more revenue for the organization.

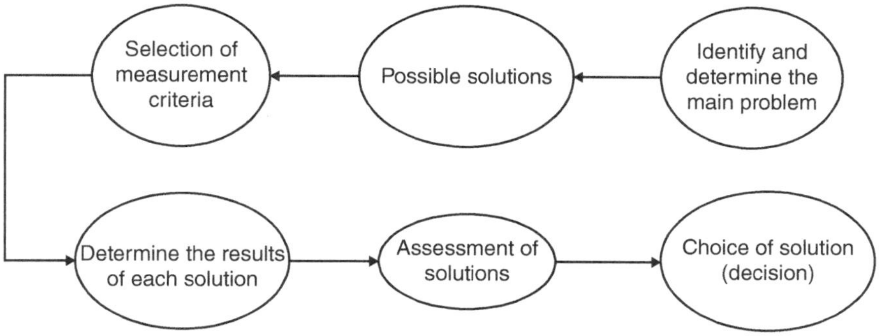

Figure 3.5 Decision making.

The fourth step is to determine the results of any of the possible solutions. At this stage, for example, the costs of using each of the solutions are calculated and based on the next assessment. Based on selected criteria, the positive and negative outcomes of each solution are determined at this stage.

The fifth step is to evaluate the solutions by examining their results. Each solution will be compared to other solutions based on the results of the organization's allocation, and the priority of the solutions will be determined. Sometimes, it is difficult to evaluate solutions and prioritize them because they are prioritized by a criterion, while they will not be prioritized for another criterion. In such cases, the manager should try to set up a compilation and make an assessment based on it.

The final stage in the DM process is to select from different ways and present a decision statement. This stage will be realized at the end of the assessment and prioritization stage, and the solution that achieves the best result and receives the highest priority level is a selective solution, but we mean the solution here as a decision– adopted by the Director, which can be considered as an independent step in the DM process. Figure 3.5 shows the different stages of the decision process.

Sometimes, the pilot implementation of the decision is also part of the DM process, in which case the feasibility of the decision is evaluated and, if successful, continues to run, and if it fails to succeed in the form of a new problem at the beginning of the DM process, the steps described above are repeated.

3.2.1 Features of a good decision

1 ***Flexibility.*** Because the conditions governing the location of the decision are variable, and each one can be influenced by different factors, the decision element should be considered as an element of flexibility and adapted to the emergencies and possible changes.

2 ***Thinking.*** In each decision, futures and information analysis play a very important role, and the decision maker must have a realistic perception of the future concerning past information and existing conditions, and to properly analyze information.
3 ***Understanding.*** The proper understanding of the aspects and characteristics of everything is very necessary, and if the understanding is not correct, expecting the correct operation also does not go. The decision maker must decide on a way that understands the underlying people with their information and understands exactly the decision maker.
4 ***Ability to establish coordination.*** If different parts of the organization seek to achieve their particular goals, then being involved with the organization as a whole of components would disrupt the goals of the organization. Therefore, since each part and work of each episode has a significant impact on other areas, the organization manager should establish a regular relationship between the different parts.

3.2.2 Types of divisions in decision making

Decisions can be divided into two main categories according to different criteria: planned and unplanned decisions, and individual and group decisions. Below is the description of these categories.

3.2.2.1 Planned and unplanned decisions

To discuss how decision makers at the top level now and in the future, they divide them into two bipolar categories: planned decisions and unplanned decisions. These two types of decisions are not separate and are a complete continuum, at the end of which there are many planned decisions, and at the other end of this continuum, there are many unplanned decisions. In the path of this conjunction, we can find various decisions that we use here, in terms of planned and unplanned, to describe the black and white spectrum of DM.

Decisions are planned to the extent that they are repetitive and conventional, as long as there is a specific way to follow them, and they do not need to be considered new every time they happen. The only obvious reason why the planned decisions tend to be repetitive or vice versa is that if a particular problem occurs, it is usually used in a conventional way to solve it. There are several examples of such planning decisions in the organization. They are determining the price of the customer's order, and calculating the amount of payment to the employee who was ill or reordered the administrative requirements.

Some decisions are not scheduled, because they are new and without a history of construction, and there is no known method for solving them because of their newness or the complex nature of their construction, although they are of such importance as to be worthy of use and have a specific solution. For example, General Eisenhower's decision to invade Germany in Europe is a good example of unplanned decisions. We do not just consider the attack command but also consider all the complex steps of obtaining information and designing operations before it. Many of the components were previously planned according to military uniforms, but before they could be designed, they needed to provide a broad range within the framework of the military policy guidelines.

The program word is taken from the computer here and is intended to be used in the same state as the one in which the string is used. A program in meaning is a version or policy **(strategy)** governing the sequential response of a system to a complex work environment. Most programs that govern the responses of the organization, such as computer programs, are not accurate or contain details. In any case, they all intend to allow the system to respond appropriately. So, in such a case, can we say that the reaction of a system to a situation is not planned? Surely, reaction is determined by something. That is a set of rules, procedures, and certainly one program. The purpose is an unplanned reaction where the system does not have a specific way of dealing with the situation: for example, the status quo; therefore, it must rely on its overall capacity for consulting, adapting, and using the problem-centric method. In addition to having specific skills and knowledge, man has a capacity and overall ability to solve problems; this situation is as new or complex as possible, and a person can root for the reasons and tools needed.

This general solution to the problem is not always workable and often fails, or results in unfavorable results. However, fewer people are in a new situation quite helpless. Is not efficient for him because he has overall problem-solving equipment, but some of the gaps and deficits, lacks in him is because his specific skills to solve the problem. Organizations, as a group of people, enjoy this overall ability to adapt.

Usually, the cost of paying for these general plans (without any prior planning) is high, and it should be used even if they do not come with these programs when there is a new issue or a secondary program for use. However, if a particular situation is created for a specific class, programs can be created for specific purposes that can be used as problem-solving tools, which provide both solutions better and costs compared to the use of the problem-solving tool generally is less than average.

The reason decisions are categorized into two categories is simply the use of different techniques that are used in different aspects of DM. This distinction makes it easier to classify these practices and is used only for that purpose; what the reader sometimes remembers is that the world is a colorless state with just a few black or white points.

Table 3.27 presentsa four-part map of the proposed coverage areas. In the upper half, the table shows how the planned decisions are related; in the lower half, the existing methods are associated with unplanned decisions, and the mid-eastern classical practices that are in the decision are used. These are a range of tools that have been used by high-level executives and organizations since the advent of history to the current generation. In the middle part of the table, new DM methods have been introduced. The tool has been continuously changed since the Second International War and has been used solely by the management.

Table 3.27 Traditional and modern decision making techniques

Modern	*Traditional*	*Type of decisions*
1 Operational research. Mathematical analysis of computer simulation models 2 Electronic data processing	1 Habit 2 Ordinary work. Standardized Operational Steps 3 Organization Structure. Normal expectations The system targets the well-defined information channels	Planned. Regular and repetitive decisions of the organization provide a specific process for its implementation.
Innovative techniques Applied to solve the problem. 1 Educate decision makers 2 Making innovative programs for computers	1 Judgment, Illumination, and Creativity 2 The rules of the fingertip 3 Selection and training of top executives	Unplanned. Abrupt decision making policies, badly regulated scenarios

3.2.2.2 Individual and group decisions

Decisions can be made individually or in groups.

Individual decisions

Decisions that managers take individually can be justified in the following ways:

1 ***The rational method.*** The rational method, i.e. the interference of system thinking in deciding in the principle of the systemic system, is to collect all relevant in-line and outsourced information. In this way, the type of effort that managers make when deciding is of interest.
2 ***The limited range of rational methods.*** The purpose of decisions that must be taken in very limited circumstances in terms of resources and time.
3 ***Rational method.*** When a person decides rationally, systematically analyzes a problem, chooses a solution, and passes logical steps one after the other. This method guides the person in DM.
4 In implementing the rational method, the DM process can be divided into eight stages.
5 ***Supervision or control over the environment.*** The manager of the internal and external information of the organization that controls deviations from the planned or acceptable behavior.
6 ***They are providing a precise definition of the type of decision.*** The manager will identify shortcomings and deviations by identifying the details of the problem.
7 ***Setting objectives.*** The manager determines what will make the decision.
8 ***Identify the problem.*** At this stage, the manager tries to find out the main cause of the problem and collect many types of data to identify the underlying cause.

9 ***They are providingsolutions.*** At this stage, the practical solutions that can be achieved through the achievement of the goals are identified.
10 ***Evaluation of Solutions.*** This step involves the use of statistical methods or personal experience in order to estimate or calculate the amount or probability of success.
11 ***They are choosing the best solution.*** This stage is the core of the DM process. The problem manager analyzes the goals and solutions that are acceptable and chooses the best solution.
12 ***Implementation of the decision.*** The manager uses his administrative and organizational capabilities and issues instructions to ensure that the decision is taken. As soon as the solution is implemented, the control stage (stage one) is restarted immediately.

3.3 Group decision

Given the rational constraints that each human being alone has and the complexity of today's modern organizations that are so wide-ranging and complicated that their management does not take on a person alone, and with the principle of always two better brain than a work brainit can be concluded that collaborative collaboration is not only a way to achieve a rational, comprehensive, and complete DM system.

Therefore, it is often necessary for major law firms to have, in addition to the director, a board of directors, and in some institutions, such as universities, the presence of a board of trustees is also required. Group DM is the mechanism by which we want to find better solutions and make more appropriate decisions. Better DM means making DM systems easier for the target, and the quality of a decision is the ultimate success rate of the decision to reach the goals and ideas of the decision maker. Group decisions, as well as individual DM, have many advantages and disadvantages, and it is the manager's task to identify the appropriate position for the optimal use of each of these two DM methods. The advantages and disadvantages of group and individual DM are relative, and the manager should design his DM system to make the most of the positive aspects of both DM possible.

Group DM is better for two reasons than individual DM:

1 The total amount of knowledge and information that is concentrated in the group is much more than the knowledge and information that exists in a single person.
2 In group DM, more and more varied solutions are proposed to solve the problem because members of the group with their own experiences regard the issue from a particular point of view, which, in turn, leads to the study of the issue of DM aspects and various angles.

However, in any case, group decision does not guarantee a high-quality decision. In terms of quality, the relative privilege of a group decision in individual decision depends on the composition of the decision group. A group decision is superior to individual DM such that the individuals who form a group have different, specialized, and experimental backgrounds, and are not the same. Also, the quality of group DM depends on the degree of interaction between the members participating in the DM group; in general, it is expected that the intergroup interaction and response among the individuals will be greater, and, as far as interactive constructive dialogue and discussion are concerned, between them, the quality of DM is also improved.

3.4 Group decision Decision Methods

Among the various types of group DM methods, here are three major issues:
(1) Crisis (Brainstorming); (2) Nominal method; (3) Delphi method.

1 ***Crisis (Brainstorming).*** The purpose of engaging or creating a brainstormin-group member is that the group tries to help those who want to agree on the members of the group and do not allow them to block the paths and obstacles to creativity and ways. New solutions.To reach this goal, the group is trying to provide an innovative solution. The front group also takes away any deterrent factor.
2 ***Nominal method.*** In this way, members of the group are not allowed to discuss during the DM process and hence call it the nominal method. All members of the group have a physical presence, but each member acts independently. Particularly, after the problem is raised, the following steps are taken:

   A Before any discussion, the members independently write their opinion on solving the problem on paper.
   B After the silence period, each member, in turn, enters the position and offers his own opinion or solution.
   C The group is discussing and evaluating comments and solutions.
   D After that, each member writes the solutions in the order of priority in silence and without consultation with others.
   E Finally, they collect the members' comments and determine the priority of the solutions. The main advantage of the methodology is that the group must be formally assembled and group members can think independently and then submit their personal opinion.

3 ***Delphi method.*** The Delphi method is similar to a nominal method, but it does not have to be physically present in the group. In this way, the members of the group are not allowed to face the call. In implementing the above method, the following steps are taken:

   A The issue is identified, and members of the group are asked to provide possible solutions to the ducts of a series of questionnaires that are accurately raised.
   B Each member completes the first questionnaire without mentioning the name and independently.
   C Results of the first collection questionnaire and answers are identified.

D A copy of these answers is given to each member.
E After each member examines the results, members are asked to submit their solutions once again. They usually provide new solutions or change their previous position.
F "D" and "E" steps are repeated to the extent necessary so that the group achieves consensus.

Why is group decision important?

The importance of group decision is mentioned below for five reasons.

The first reason is the issue of legitimacy. If a person decides alone, it may be deemed by others to be arbitrarily adopted without respect for the interests and feelings of others, and enforcement is compulsory, rather than merely based on satisfaction. If the manager does so, then the decision may be questioned in terms of legitimacy. However, if the group approves the decision, it will be legitimate for the people to be approved.

The second reason is the quality of the decision. Although the group may make an inappropriate decision, it is expected that the overall quality of the decisions made by the group will be better, given the available reasons. Based on sensory (intuitive) and research studies, one can safely predict that a group can offer more than one person. The group will be more aware of potential problems and obstacles in terms of experiences, abilities, and expertise.

The third reason is innovation. In situations where the organization needs new ideas or problems require innovative solutions, it is understood that the group's situation and position are better than the individual. As the organization needs new ideas, the group is considered to be a more meaningful resource. In order to benefit from the group's privileges, techniques such as "brainstorm" have been developed and expanded, widely used in situations where initiative and creativity are needed.

The fourth reason is the lack of information. When an organization faces a lack of information, the use of a group for DM provides more information. By gathering a group of informed people, access to the most relevant and complete information becomes easier. The group cannot always have all the information, but it can help more about what information is available and what information is not available.

The fifth reason is morale. Many research studies show that participation in group DM with the strengthening of morale and job satisfaction is directly related to issues such as job stress and job satisfaction. In groups, not only are individuals socially interacting, but a role in DM can, in addition to providing positions for self-expression and personal promotion, enhance the prestige and credibility of the individual.

In group DM, J. Hall's research suggests that group DM quality will be better when the group members follow the guidelines below:

1 Point of view is as rational as possible.
2 When the debate reaches the point of silence, one should not be convinced that anyone is a winner or a loser, but should seek acceptable solutions for all interest groups.

3 In order to avoid conflicts and to reach an agreement and early coordination, members of the group should not change their opinions quickly.
4 Contradictory techniques such as majority voting and bargaining should be avoided.
5 Differences of beliefs are natural and can be expected, and opposition contributes to group DM as it brings the group together with a range of information and ideas, and provides a better opportunity to the group to reach an appropriate solution.

Under what circumstances should decisions be made individually?

Although group DM has many effects on the organization, it is more appropriate in some situations if decisions are taken individually. These conditions can be recognized by the following factors:

1 ***Urgency.*** In some cases, where the issue is urgent and time-bound, DM through the group's deliberation is not desirable. So, the time limit prevents people from being involved in the decision.
2 ***Unique Knowledge.*** In cases where solving the problem involves the use of specialist knowledge, group DM is not a proper technique.
3 ***Confidentiality.*** When information is confidential and secret, it is better for the person who has access to the information to decide and thus avoid disclosure of confidential and secret material.

What passed an overview of the methods of decision making? According to the above, competent authorities in different organizations, through the use of feedback, experience, and expertise of others, improve the quality of decision making in organizations.

3.5 Major Decision Making Models

A model is a pattern taken from reality and shows the relationships between the variables and can be used to predict the decision. The decision maker can create a model of the system and then examine it with different results from various decisions. By adopting the model, without making a real risk of real-world DM, we can make the most desirable decision. In most cases, it is impossible or difficult to effectively test the decision and choose the best of them, while this is easily possible using the model. By a model, the effects of different changes can be measured quickly and accurately, and the decision maker will be informed of their outcomes without risk or with risk.

The main problem in modeling is that it is sometimes not a realistic model and does not correctly represent the original system. In such a case, the conclusion and reliance on the model would not be helpful and would be misleading. When designing, it should answer the question of whether the model has all the components of the actual system. Moreover, do these components represent facts?

Knowing the modeler of the system and the perspective of informed people can help him in making a true and fair model. Also, the model designer can match the results with real functions and ensure the accuracy and predictive power of the test stage. Of course, it should be kept in mind that all categories of decision cannot be represented in the model, and only the recognition of the components of a system and their relationships can be expressed in the design of the payment model.

Models are an understanding of reality, but they are not the same reality and therefore do not represent the exact systems of reality. Of course, this is not a negative feature of the model, because one of the objectives of the model is to simplify and show the main components of the system and sometimes in the modeling of additional and disruptive factors, they are deliberately set aside to obtain a model showing Providing the main components, and the relationship between them is simple enough and easy to use.The optimal model is a model that has the main components of analysis and DM, and although it is not just as real, it provides an analyst and decision maker with a simple and suitable means by showing the relationships between the components. Below are some basic DM models.

3.5.1 Satisfactory Model

The concept of this model is that decision makers encounter a complex problem because they are not able to gather all the information, need to understand optimization for the ability of the human intelligence process, and are coming to reduce the problem to an understandable level. Because the human thought's capacity for formulating and solving complicated problems is too small, it cannot, therefore, be subject to the conditions of complete rationality. Therefore, the person acts in the framework of limited rationality. They create a simplified model to get the complexity of their faces and displays. Then, one can show reasonable behavior within the limits of this simplified model. In this method, the investigation of options begins when a problem is detected. Because the list of criteria is likely to be limited and does not respect all aspects, a limited list of visible choices will be provided. These are usually options that are easy to find and are greatly appreciated. Often, these are familiar criteria that have already been tested and are real solutions. When this number of limited options was specified. The decision maker will begin to review them. But this review does not have a compression state, which means that all options are not evaluated. Instead, the decision maker starts evaluating options that are slightly different from the current solution. By moving on familiar and appropriate routes, the decision maker continues to evaluate the options, until the satisfactory option is specified. Then the decision maker chooses the solution as good enough and does not continue to research about obtaining an optimal solution. The first option that matches the "good enough" criterion ends the search, and then the decision maker can try to implement this solution.

One interesting aspect of the satisfactory model is that the order in which the options are evaluated is important in knowing which option is selected. If the decision maker seeks to optimize, then all options are categorized in priority order because they are fully explored, and therefore the recommendation of each of the potential solutions will be subjected to a complete assessment. However, this is not true about the satisfactory model. Suppose a problem has more than one potential solution. The satisfactory choice in the initial decision maker's encounter will be the first acceptable method. Because decision makers use simple and restrictive models, they usually choose evident options, that is, those who are familiar with them or who are not in many ways in conflict with the current situation. Those solutions that are at least separating from the existing situation and complying with the decision criteria are the most likely to be chosen. This suggests why many of the decisions that people make do not lead to choices completely different from what they have already found. A unique option may be an optimal solution to a problem that is rarely chosen in any case. An acceptable solution will be selected before the decision maker is required to investigate a solution other than the status quo.

3.5.2 Implicit Interest Model

Another model that is designed for complex and unusual decisions is the implicit interest model. This model, like the satisfying model, raises the argument that a person solves complex problems by simplifying processes. In any case, the simplification of the implicit interest model implies that it should not be entered into the DM process in a rigorous stage until one of the options can be identified as implicit interest. In other words, the decision maker is neither rational nor realistic. Then, the rest of the DM process will necessarily be a practice to confirm the decision; in other words, the decision maker seeks to ensure that the option of his implied interests is, in fact, an appropriate option.

The implicit interest is the result of research graduate students; the university is M.I.T.[1] They knew this model openly, and for many years, they used it for solving organizational problems and case analysis (CASES) in accounting, finance, management, marketing, and quantitative methods. Besides, the decision to choose a job was one of the important decisions. If the decision was made, and it was supposed to be used with the optimization model, and the teams had some experience in using it, they would have to use it. However, research has shown that the optimization model is not working. However, the implicit interest model is also an accurate description of the actual DM process.

Once the problem has been identified, the decision maker chooses his first choice but does not end the research at this point. The decision maker is often unaware of the fact that he has already indicated his tacit interest, and the rest of the way, in the sense of a recipe for exercising that is biased, so many options are available. This is very important because it gives objectivity. Then the approval process begins, and the options are reduced to two categories: the candidate (candidate) and the candidate. If the nominee is the only suitable way for growth, the decision maker will try to obtain an acceptable option to become an appointive candidate, and thus, the decision maker will gain something for comparison. At this point, the decision maker makes weighting factors for DM; of course, many distortions of interpretation are also made. The relevant weighting factors, after selecting and shaping, will ensure that the success of the option will be of interest. Of course, this is exactly what will be achieved. The explicit evaluation shows the candidate's superiority to the approved candidate.

If the implicit interest model is used, the research will seek to find options before the decision maker wants to say that he has decided. M.I.T students were able to accurately predict employment in research that 87% of professional jobs can be accepted by them before they confirmed that they have reached the final decision. The notion of speech is that the DM process is more influenced by the inspirational feeling than the rational reality.

3.5.3 Maximum Use Model (Optimization)

First, let us look at how a person should behave in order to maximize his returns. We call this action in making maximum model decisions.

Steps to maximize the use of the model (optimization)

The six steps that a person must follow explicitly or implicitly when deciding aresummarized in Table 3.28.

Step One. Need to make a decision. In the first step, you need to decide on the necessity. As previously stated, the difference between the state and the status quo means the existence of a problem, which must be decided to resolve. If one calculates his monthly expenses and finds that their value has increased from his income, then he will surely have to make a proper decision to solve such a problem since the difference between the cost and the actual cost is significant.

Step Two. Identify decision criteria. When a person realizes the need for a decision, he must identify the criteria and identify who is important in the DM process. For example, consider a high school student who faces the issue of

Table 3.28 Steps to maximize the use of the model (optimization)

Steps to maximize the use of the model (optimization)	*Row*
Ensuring the need for decision	1
Determine decision criteria	2
Determine the weighting factors of the criteria	3
Increase options	4
Evaluate options	5
Choose the best option	6

choosing a college to study. This student can add the following factors to their list of criteria:

Schools that may have their friends, part-time work, and whether the first-year student can use a college dormitory. Another student who wants to go on to study and decide to go to college will probably consider other criteria.

If this is not a standard list, then we think it is not critical to the decision maker.

Step Three. Determine the coefficient of each criterion. The criteria set out in the previous step are not all of the same importance, so each of the criteria listed in the second step must be multiplied or weighed to determine their priority and importance in DM. Although all these criteria are important, some of them are even more important.

How does the decision maker give weight or coefficient to each of these criteria? The easiest way is to write one (for example, 10) against the most important criterion, and then, according to this standard, the coefficient of the other criterion is also determined. Therefore, the result is that the decision maker acts in the second and third steps according to personal taste.

Step Four. Identify solutions. In the fourth step, the decision maker should determine how this problem can be resolvedsuccessfully. At this stage, it is only necessary to list these routes and does not need to evaluate them.

Step Five. Assess ways or modes. Once solutions are identified, the decision maker must carefully evaluate each one of them and compare them with the criteria and coefficients given in steps two and three.

Step Six. Choose the best solution. In the optimization pattern, the last step is to choose the best solution (among the ways, modes, or solutions that have been evaluated). Because the solution that brings the highest score is the best, so choosing the best solution is not a problem at all. The decision maker chooses the only solution, which is the fifth step has the highest score.

3.5.3.1 Maximum utilization Model Assumptions (Optimization)

The steps taken in the optimization pattern include several assumptions. If we have to look at how well the optimization model represents individual decisions, we must understand these assumptions.

The assumptions of the optimization pattern are just as rigorous as the assumptions are. In rationalism, or rigorously and sensibly, solutions are chosen that have the highest value and stability. Therefore, making a sound decision means that the decision maker follows rational principles and does not in any way apply his personal opinion. It is assumed that the person has a clear goal and the six stages of the optimization pattern lead him to choose the best solution. Hence, the goal is of the highest value. Below are the assumptions that are madewisely.

Goal-oriented. In the optimization model, it is assumed that there is no conflict in terms of goal determination. This decision may concern the selection of the faculty, whether the presence or absence of the work or the selection of a qualified candidate for an organizational vacancy or something else. It is assumed that the decision maker always has a clearly defined goal of trying to maximize it.

All roads are known. It is assumed that the decision maker can identify all relevant criteria and list all the states and solutions. The optimization model gives a picture of the decision maker that he is fully capable of measuring various criteria and solutions.

Priorities are clear. In a costly model, it is assumed that the criteria can be scored and ranked according to priority.

Priorities are stable. Since the purpose and priorities are clear, it is assumed that the criteria belonging to a particular decision are fixed, and the weights or the coefficients which are given to them will not change over time. Therefore, these criteria and solutions can be achieved at any time.

The final option yields the most results. The person who implements the optimization model and makes the logical decision will choose the solution that has the highest score. According to the sixth step, the best solution will be the most profitable.

3.5.3.2 Model-based Optimization Forecasts

According to the previous assumptions, one can predict that the decision maker is in such a situation. A clear goal, a set of criteria that determine the relevant factors in that decision, the priorities, and the order of the criteria are clear. And over time, this arrangement does not overlap, and ultimately, the decision maker chooses the one that has scored the highest after examining all the solutions, and since all the information obtained is in a fair and stable way, Have been evaluated, leaving no room for concern to the decision maker. In Figure 3.6, the optimization process steps are summarized.

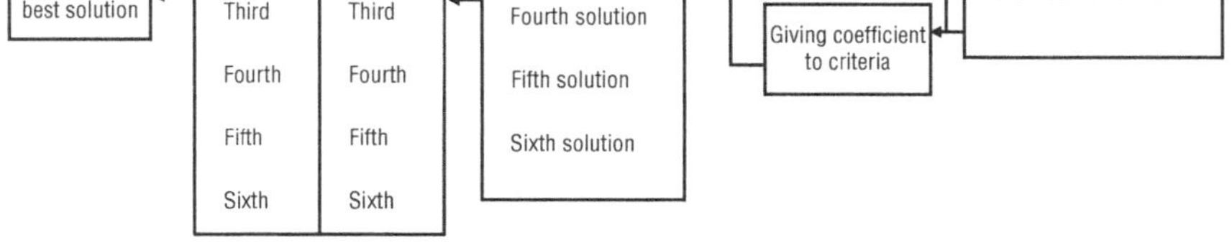

Figure 3.6 Optimization model.

3.6 Large Decision Making Environments

Major DM environments are listed below.

3.6.1 Definite

In this case, all the required DM information is available, and we know that decision A leads to I. The techniques of this kind of DM are divided into two categories: inefficient techniques and interactive techniques. In non-interactive techniques, the strength of a benchmark cannot compensate for other benchmark weaknesses, but not in interactive techniques. The most popular non-MADM[2] techniques are:

Dominance, Pessimistic, Optimistic, Conjunctive-Satisfying Method, Disjunctive-Satisfying Method, Ordinal, Cardinal (SAW, TOPSIS, ELECTRE, AHP, WP, HRM), Lexicographic, Elimination, Compensatory Methods, Non-Compensatory Methods. The most famous are ELECTRE, TOPSIS, and AHP techniques. Each of these methods is further defined and reviewed.

3.6.2 Under the risk

In a risk-taking decision, a particular action may lead to different consequences. Here, it is assumed that the probability distribution function is known to occur. In DM, we have a decision maker on the one hand that has a set of actions or decisions, and on the other hand, we have nature that can create different situations with different possibilities for the decision maker. Risk-solving techniques are as follows:

1 Naive techniques
2 Priori techniques
3 Posteriori techniques

3.6.3 Uncertainty

In this case, performing a particular action leads to different results than the distribution probability of their occurrence is not known. We use "game theory"[3] to decide on the uncertainty conditions. The theory of games is a mathematical theory that deals with competitive situations. This theory is useful when two or more individuals or organizations try to make decisions with conflicting goals. In such a situation, the decision of one of the decision makers will affect the rest of the decisions.

3.7 Define MCDM[4] and its basic concepts

Definition of multiple-criteria decision making

Models of optimization from the era of the industrial movement in the world, and especially since World War II, have always been the subject of interest by mathematicians and industry practitioners. The main emphasis in classical optimization models is to have a measurement criterion (or a target function) as follows (use the following method to optimize):

$$f(x); f: E^n \to E^1 \qquad (3.23)$$

$$s.t: g_i(x)\begin{bmatrix} \leq \\ \geq \\ = \end{bmatrix} \circ; i = 1, 2, \ldots, m; \quad E^n \to E^m$$

So that the model can be linear, nonlinear, or mixed altogether, but in recent decades, attention has been focused on multi-criteria models (MCDM) for complex decisions. In these decisions, instead of using an optimality measure, several measurement criteria may be used.

These decision models are divided into two major categories: multi-objective models (MODM[5]) and multi-index models (MADM) such that multi-objective models are used for design, while multi-index models are used to select the best option.

Multi-objective model (MODM) can be formulated as follows:

$$\{f_1(x), f_2(x), \ldots f_k(x)\} = F(x) \qquad (3.24)$$

$$s.t: g_i(x)\begin{bmatrix} \leq \\ \geq \\ = \end{bmatrix} \circ; i = 1, 2, \ldots, m$$

$$x \in E^n$$

This model is famous for VMP,[6] and the design of the optimum point for it from a non-empty set $(S \in E^n)S$ will be available.

Multi-index model (MADM) is formulated as a decision matrix.

$A_m, \ldots, A_2, A_1$. In the decision matrix D, the prefix m of the option m is already known (such as buying a plane from a particular manufacturer), $x_n, \ldots, x_2, x_1$ indicates n indicators (or attributes) such as cost, capacity, profitability, convenience, reputation, etc., to measure the desirability of each option, and eventually elements r_{ij} specifies the specific values of the j index for option i.

Indicators may be a bit (like cost) or qualitative (such as convenience).

3.8 Definitions and basic concepts

Objective as a function. This is the proper mathematical direction for optimization, the directions that DM[7] needs to design to optimize its decision.

Optimal solution. A solution that optimizes for each of the target functions (or for each of the indicators) simultaneously. For example, for a multi-objective maximum problem$^*X$, an optimal solution for VMP would be if and if. For all, of course, the optimal solution for most situations for a VMP will not be due to conflicts between goals. An optimal solution for a multi-index decision (MADM) will include the most suitable hypothetical option A^*, to have.

$$A^* \approx \left\{x_1^*, x_2^*, \ldots x_n^*\right\} \tag{3.25}$$
$$x_j^* = \max u_j\left(r_{ij}\right); i = 1, 2, \ldots, m$$

So $uj(\circ)$ the utility of the j index is, in other words, A^*. It consists of the preferred value or desirability of any existing indicator from the decision matrix, which, of course, is such an ideal option (A^*). In most cases, there will be no external DM.

Alternative. One of the suggested solutions.

Criterion. A feature that needs to be in place to be acceptable.

Decision matrix. A matrix in which the value of each criterion is recorded for each alternative.

Preferred solution. The solution chosen by DM is through effective (or non-dominant) solutions using other mental criteria.

Satisfying solution. A solution that has predetermined goals for making decisions more than needed. These solutions may not be effective solutions, but their simplicity is consistent with DM's behavior.

Effective solution (non-dominant). An effective solution will be if the existing values of all objectives cannot be improved simultaneously by other practical means. In other words, an effective solution would be if there were no other practical remedies available for that.

$$\left\{\begin{array}{l} f_i(x) \geq f_i(\tilde{x}); \text{For all } i \\ f_i(x) > f_j(\tilde{x}); \text{At least for one } j \end{array}\right\} \tag{3.26}$$

Stages of DM matrix preparation.

There are a few important points in the decision matrix that are as follows:

A Some criteria are quantitative and measurable if some of the other criteria are qualitative.
B The criteria do not have the same dimension. Therefore, it is not possible to perform mathematical exercises on them.
C Criteria do not necessarily have the same importance in DM.

Considering the topics mentioned earlier, DM techniques require that the qualitative criteria in the matrix be converted to small values and that the dimensions of all the criteria are the same. Meanwhile, methods should be defined to determine the weight of each criterion or its importance in DM. To make such changes in the matrix, the following applies.

3.9 Qualitative Criteria to Become a Little

In order to convert qualitative criteria to a certain extent, we can use scale-level lines or fuzzy logic method.[8]

3.9.1 Scale ruler

In this method, they assign a maximum and minimum quantitative, qualitative value and calibrate the values between these two points. Figure 3.7 shows an example of a scale ruler. Due to the limitations of this method, it is used less today.

3.9.2 Fuzzy logic

The most effective way to convert qualitative criteria is to use fuzzy logic slightly. The foundation and basis of fuzzy logic were founded in 1965 by presenting a paper titled "Fictitious Collections" by Dr. Lotfizadeh, a professor of Iranian-language studies at the University of California, Berkeley. In fuzzy logic, the correctness or inaccuracy of anything is relative, in contrast to the binary logic (Boolean algebra[9]) in which logical operations have two values 0 or 1 or true to the wrong. Binary logic is a subset of fuzzy logic. In fuzzy logic, the membership function has a degree between 0 and 1, and instead of using one of the members of the set {1 and 0}, members are used to indicating membership from 0 to 1...

For example, the young fuzzy subset is displayed as follows (Figure 3.8):

$$\text{Yong}(x) = \begin{Bmatrix} 1, \text{ if } \text{age}(x) \leq 20; (30 - \text{age}(x) / 10, \\ \text{if } 20 \prec \text{age}(x) \prec 30; 0. \text{ if } \text{age}(x) \geq 30 \end{Bmatrix} \tag{3.27}$$

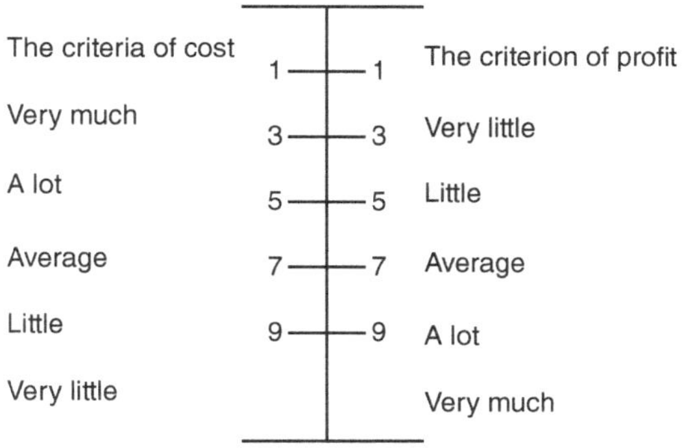

Figure 3.7 Scale ruler.

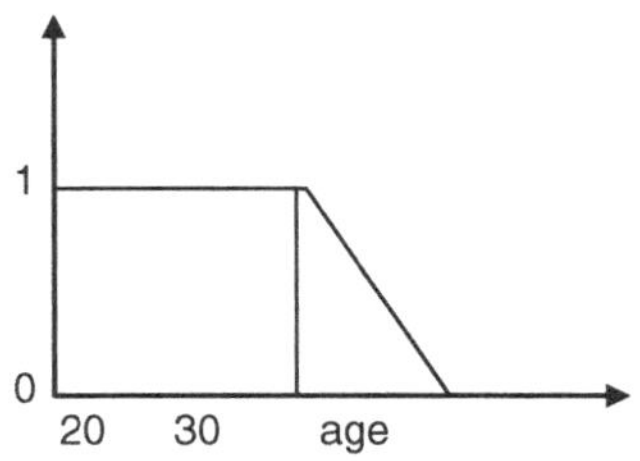

Figure 3.8 Young fuzzy.

3.10 Normalize

After converting the quantitative criteria to a bit, it is time to equalize the dimensions of the criteria. This unification means meaningless dimensionless metrics. This process is called normalization. There are several methods to normalize, and each method of DM offers its method.

3.10.1 The first Method of Normalization (Vector)

This normalization method maintains the numbers of profits from the profit and the cost-related numbers of the cost.

$$r_{ij} = \frac{x_{ij}}{\sqrt{\sum_{i=1}^{m} \left(x_{ij}\right)^2}} \tag{3.28}$$

3.10.2 The second Method of Normalization (Linear Normalization)

In this method, after normalizing, and analyzing the criteria for the profit we need to do the same action related to costs, as you can see in (3.10.2.2). This method offers different formulations for the profit and cost criteria; the roof is one, but there is no specific amount for the floor.

3.10.2.1 Profit Criterion

$$r_{ij} = \frac{x_{ij}}{x_j^*} \; x_j^* = \max\left\{x_{ij}\right\} \tag{3.29}$$

$$i = 1, \ldots, m$$

3.10.2.2 Criterion of Cost

$$r_{ij} = \frac{x_j^{\min}}{x_{ij}} \; x_j^{\min} = \min\left\{x_{ij}\right\} \tag{3.30}$$

$$i = 1, \ldots, m$$

The weakness of this method is that it does not have a specific number for the most undesirable value.

3.10.3 The third Method of Normalization

This method does not have a specific name and gives the most favorable criterion number one and the least favorable criterion a value of zero.

3.10.3.1 Profit Criterion

$$r_{ij} = \frac{x_{ij} - x_j^{\min}}{x_j^* - x_j^{\min}} \tag{3.31}$$

3.10.3.2 Criterion of Cost

$$r_{ij} = \frac{x_j^* - x_{ij}}{x_j^* - x_j^{\min}} \tag{3.32}$$

3.10.4 The fourth Method of Normalization (Quadratic Norm)

If data processing is based on the probability distribution of the function, the norm scalar will be the norm input of the quadratic. In a quadratic norm, the data is converted into a relative frequency distribution on any scale, such as the Shannon entropy technique.

$$n = \frac{r_{ij}}{\sum_{i=1}^{m} r_{ij}}; \forall_j \tag{3.33}$$

3.10.5 The fifth Method of Normalization (Euclidean Norm)

If the logic of data processing is of the type Euclidean, that is, on the distances of the Euclidean (the torque of the second order), then the norm of the non-scaling of the data is Euclidean norm, such as TOPSIS.

$$n_{ij} = \frac{r_{ij}}{\left[\sum_{i=1}^{m} r_{ij}^2\right]^{\frac{1}{2}}}; \forall_j \tag{3.34}$$

3.10.6 The sixth Method of Normalization (Zero and One)

Norm Z can be applied if all of the criteria of the D matrix are normal and apply to techniques that are assumed to be normality standards.

$$Z_{ij} = \frac{r_{ij} - M_i}{\delta_i}; \forall_j \tag{3.35}$$

3.11 Weighting

Now, it is time to determine the importance of (weight) criteria. To do this, use the following two methods:

3.11.1 Shannon Entropy Method

In this method, the following steps are taken:

A By using the following equation, normalize the decision matrix. It should be noted that the use of this method involves the transformation of qualitative criteria into quantitative.

$$r_{ij} = \frac{x_{ij}}{\sum_{i=1}^{m} x_{ij}}; \forall_{i,j} \tag{3.36}$$

B For the *j* criterion, calculate the entropy using the below formula:

$$E_j = -K\sum_{i=1}^{m} r_{ij} \times 1nr_{ij}, \quad K = \frac{1}{1n\ m} \circ \quad \leq E_j \leq 1 \tag{3.37}$$

C For each *j* criterion, calculate d_j.

$$d_j = 1 - E_j \tag{3.38}$$

D Determine the weight of the *j* criterion.

$$W_j = \frac{d_j}{\sum_{j=1}^{n} d_j} \tag{3.39}$$

3.11.2 Paired Comparison Method

Another common method for calculating the weight of the criteria is the paired comparison method. This procedure works in the following order:

A A square matrix whose rows and columns are metrics.
B We ask the decision maker to compare the importance of each of them to a number between 1 and 9 by comparing the two criteria and to enter the values in the matrix.
C The resultant matrix will be a matrix whose original diameter is 1, and all elements are equal to the original diameter of the image. At this point, we normalize the matrix. To do this, we divide each element in the matrix into the sum of the elements of the corresponding column.
D The average of each normalized matrix line is calculated. The values obtained will be the weight of the relevant criteria.

3.12 Types of MCDM Techniques

MCDM techniques are divided into two categories, interactive techniques and non-intrusive techniques, which are described below.

3.12.1 Non-Interactive (Inventive) Techniques

Inefficient techniques, the strength of a benchmark, cannot compensate for other benchmark weaknesses. Since these techniques are very simple and at the same time provide a good approximation of the chosen alternative, they are also referred to as innovative techniques. The most popular techniques are:

- Dominance
- Maximin
- Maximax

- Conjunctive
- Disjunctive
- Lexicography
- Elimination

Dominance. In this alternative, a superior alternative is an alternative that is better or equal to all other criteria. This technique does not require any initial preparation of the decision matrix, but usually, there is less use for it.

Maximin. Maximin's principle is based on the fact that the resistance of a chain only depends on the strength of its weakest loop. In such situations, the function of an alternative depends only on its weakest criterion; it should be chosen as the alternative, which is the best alternative in terms of this weakest criterion. To apply this method, the DM matrix must be normalized first. To this end, a technique must be used to first classify the criteria of the cost of the sex as a profit, and second, to define the limit for the worst of each criterion.

Maximax. In the Maximax method, an alternative is chosen which, in its strongest position, is superior to other alternatives.

Conjunctive. In the Conjunctive technique, for each criterion, a standard limit is defined. Alternatives that are above or below this standard are accepted, and the rest are discarded. This approach may not lead to a single solution at all times, but it can be used to minimize the decision space and eliminate undesirable alternatives.

Suppose that we have a set of n independent variables and that all have the same weight. If r is the part of the alternative that should be laid out and P_c is likely to have an alternate selected randomly above the standard, then we will have

$$P_c = (1-r)^{1/n} \tag{3.40}$$

Disjunctive. In this method, for each criterion, an ideal limit is defined, and the alternative is achieved in one of its criteria to this ideal limit.

Lexicography. In some DM processes, it seems that a benchmark is a predominant criterion. In this technique, the criteria are ranked according to the importance of the decision maker; then, based on this ranking, the best alternatives are selected, so that we first consider the criterion of the priority, and then the alternatives that are superior to this criterion which we determine.

Elimination. If at this stage the answer is the same, we stop and say the same as the optimal answer. Otherwise, if the answer is not the same, then the criterion is chosen with the second priority, and we select from the alternatives selected in the previous step, which is the second highest criterion. If the answer is not unique, the solution will be repeated.

Elimination. In this technique, unlike previous techniques directly aimed at finding the best alternatives, this is done indirectly, so that step by step following different criteria of the undesirable alternatives are eliminated, and gradually the siege ring gets around the best of them to make alternatives.

3.12.2 Interactive Techniques

In these techniques, the strength of a criterion can compensate for other benchmark weaknesses. The most popular techniques are:

- SAW
- TOPSIS
- ELECTRE
- AHP
- DEMATEL
- NALADE

SAW.[10] Simple Additive Weighting (SAW), which is also known as a weighted linear combination or scoring methods, is a simple and most often used multi-attribute decision technique. The method is based on the weighted average. An evaluation score is calculated for each alternative by multiplying the scaled value given to the alternative of that attribute with the weights of relative importance directly assigned by decision maker followed by summing of the products for all criteria. The advantage of this method is that it is a proportional linear transformation of the raw data which means that the relative order of magnitude of the standardized scores remains equal.

This method is one of the oldest methods used in MCDM such that with the assumption of the vector W (weights of the importance of the indexes), the most suitable option is (A^*)calculated as follows:

$$A^* = \left\{ A_t \middle| \max \frac{\sum_j w_j, r_{ij}}{\sum_j w_j} \right\} \tag{3.41}$$

Moreover, if $\sum_j w_j = 1$, we will have

$$A^* = \left\{ A_i \mid \max \sum_j w_J \cdot r_{ij} \right\} \tag{3.42}$$

SAW technique is one of the most used MADM techniques. It is simple and is the basis of most MADM techniques such as AHP and PROMETHEE[11] that benefit from the additive property for calculating the final score of alternatives. In the SAW technique, the final score of each alternative is calculated as follows, and they are ranked.

$$P_i = \sum_{j=1}^{k} w_j \cdot r_{ij; \, i=1,2,\ldots,m} \tag{3.43}$$

where r_{ij} are normalized values of decision matrix elements and calculated as follows:

For-profit attributes, we have

$$r_{ij} = \frac{d_{ij}}{d_j^{\text{Max}}}; d_j^{\text{Max}} = \underset{1 \le i \le m}{\text{Max}} d_{ij}; j = 1, 2, \ldots, k \tag{3.44}$$

$$r_{ij} = \frac{d_j^{\text{Min}}}{d_{ij}}; d_j^{\text{Min}} = \underset{1 \le i \le m}{\text{Min}} d_{ij}; j = 1, 2, \ldots, k \tag{3.45}$$

If there is any qualitative attribute, then we can use some methods for transforming qualitative variables into quantitative ones.

TOPSIS.[12] In this method, in addition to considering the distance of an option from an ideal point, its distance from the ideal point is also considered. It means that the choice option should have the least distance from the ideal solution and at the same time has the furthest distance from the ideal negative solution.

The underlying facts of this method are as follows:

A The desirability of each index must be uniformly incremental (or decreasing) (the higher, the greater, the greater the utility or the reverse), which is the best value of an ideal index representing the worst value out of it the ideal will be negative.

B The distance of an ideal option (or negative ideal) may be calculated as the Euclidean distance (second power) or as the sum of absolute magnitudes from linear intervals (known as block intervals), depending on the exchange rate and there is an alternative between the indices.

The TOPSIS method was first developed by Hwang and Yoon (1981) and ranks the alternatives according to their distances from the ideal and the negative ideal solution, i.e. the best alternative has simultaneously the shortest distance from the ideal solution and the farthest distance from the negative ideal solution. The ideal solution is identified with a hypothetical alternative that has the best values for all considered criteria, whereas the negative ideal solution is identified with a hypothetical alternative that has the worst criteria values. In practice, TOPSIS has been successfully applied to solve selection/evaluation problems with a finite number of alternatives because it is intuitive and easy to understand and implement. Furthermore, TOPSIS has a sound logic that represents the rationale of human choice and has been proved to be one of the best methods in addressing the issue of rank reversal. In this chapter, we extended TOPSIS for KM[13] strategies selection problem because of the following reasons and advantages as Shih and his cooperators did for consultant selection problem.

A A sound logic that represents the rational of human choice.

B A scalar value that accounts for both the best and worst alternatives simultaneously.

C A simple computation process that can be easily programmed into a spreadsheet.

D The performance measures of all alternatives on attributes can be visualized on a polyhedron, at least for any two dimensions.

3.12.2.1 Common Methods of Normalization for TOPSIS

For MADM, a decision matrix is usually required before the beginning of the process. The decision matrix contains competitive alternatives row-wise, with their attributes' ratings. Normalization is an operation to make these scores conform to or reduce to a norm or standard. To compare the alternatives on each attribute, the normalized process is usually made column-wise, and the normalized value will be a positive value between 0 and 1. In this way, computational problems, resulting from different measurements in the decision matrix, are eliminated. Attributes have been partitioned into three groups: benefit attributes, cost attributes, and non-monotonic attributes. A few common normalization methods are organized in Table 3.29. These are classified as vector normalization, linear normalization, and fuzzy normalization to fit real-world situations under different circumstances. Additionally, three forms for linear normalization are listed in Table 3.29.

Table 3.29 Common normalization methods

Vector normalization	(1) $r_{ij} = \frac{x_{ij}}{\sqrt{\sum_{j=1}^{m} x_{ij}^2}}; j = 1,2,\ldots,m;\ i = 1,2,\ldots,n$ (2) $r_{ij} = \frac{x_{ij}}{\sqrt{\sum_{j=1}^{m} x_{ij}}}; j = 1,2,\ldots,m;\ i = 1,2,\ldots,n$	(3.46)
Linear normalization	(1) $r_{ij} = \frac{x_{ij}}{x_j^*}; j = 1,2,\ldots,m;\ i = 1,2,\ldots,n;$ $x_j^* = \max_i (x_{ij})$ for benefit attribut (2) $r_{ij} = \frac{\tilde{x}_j}{x_{ij}}; j = 1,2,\ldots,m;\ i = 1,2,\ldots,n;$ $\tilde{x}_j = \min_i (x_{ij})$ for cost attribut	(3.47)

Fuzzy Normalization	(1) $r_{ij}=\frac{x_{ij}-\tilde{x}_j}{x_j^*-\tilde{x}_j}; j=1,2,\ldots,m;$ $i=1,2,\ldots,n;$ for benefit attribut (2) $r_{ij}=\frac{x_j^*-x_{ij}}{x_i^*-\tilde{x}_i}; j=1,2,\ldots,m;$ $i=1,2,\ldots,n;$ for cost attribut (3.48)

3.12.2.2 Normal Distribution

In many applications in which some random variable X is normally distributed with mean μ and variance s^2, we will standardize X to obtain z-scores ($z=(x-m)/s^2$). The distribution of the z-scores is the standard normal distribution, that is, the normal distribution with a mean of 0 and a variance of 1.

Therefore, if X complies $N(m, s^2)$, then z abides by $N(0,1)$ also. The probability density function of the standard normal distribution is as follows:

$$f(x)=\frac{1}{\sqrt{2\pi\sigma^2}}e^{\frac{-(x-\mu)^2}{2\sigma^2}} \quad (3.49)$$

The cumulative distribution function (CDF) of a probability distribution contains the probabilities that a random variable X is less than or equal to X. The CDF of the normal distribution is expressed as follows:

$$\varphi(z)=\int_{-\infty}^{z}\frac{1}{\sqrt{2\pi\sigma^2}}e^{\frac{-(x-\mu)^2}{2\sigma^2}} \quad (3.50)$$

The normal distribution is considered the most prominent probability distribution in statistics. There are several reasons for this. One of them is that normal distribution is very tractable analytically, that is, a large number of results involving this distribution can be derived in an explicit form.

ELECTRE.[14] In this way, instead of ranking options, a new concept known as the "non-ranked" is used that way, for example, $A_k \rightarrow A_l$ suggests that although options 1, k have no mathematical preference, DM and the analyst accept the risk of A_k being better off.

In this method, all options are evaluated using non-ranked comparisons and eliminated by ineffective options.

Paired comparisons based on the degree of agreement between the weights and the degree of difference between the values of the weightings (V_{ij}) were built and tested together to evaluate their options.

All of these processes are based on a coherent set and an inconsistent set, which is also known as "coordinate analysis."

AHP.[15] This method is based on human brain analysis for complex and fuzzy problems. Since the comprehension of complex phenomena and complex problems for the human mind can be problematic, the decomposition of a big problem into its elemental elements (using a class structure) can help to understand human beings. In this method, the relationship of each element with other elements must be specified in a class structure and at different levels, and the relevance of the main objective of the problem to the lowest of the existing class of hierarchies is precisely clarified. In the structure of AHP, each element of a given level is dominated by some or all of the elements on the surface immediately above its own.

The AHP, introduced by Thomas Saaty (1980), is an effective tool for dealing with complex DM and may aid the decision maker to set priorities and make the best decision. By reducing complex decisions to a series of pairwise comparisons, and then synthesizing the results, the AHP helps to capture both subjective and objective aspects of a decision. Besides, the AHP incorporates a useful technique for checking the consistency of the decision maker's evaluations, thus reducing the bias in the DM process.

How the AHP works?

The AHP considers a set of evaluation criteria and a set of alternative options among which the best decision is to be made. It is important to note that, since some of the criteria could be contrasting, it is not true in general that the best option is the one which optimizes every single criterion, rather than the one which achieves the most suitable trade-off among the different criteria.

The AHP generates a weight for each evaluation criterion according to the decision maker's pairwise comparisons of the criteria. The higher the weight, the more important the corresponding criterion. Next, for a fixed criterion, the AHP assigns a score to each option according to the decision maker's pairwise comparisons of the options based on that criterion. The higher the score, the better the performance of the option concerning the considered criterion. Finally, the AHP combines the criteria weights and the options scores, thus determining a global score for each option, and a consequent ranking. The global score for a given option is a weighted sum of the scores which is obtainedconcerning all the criteria.

3.12.3 Features of the analytic hierarchy process

The AHP is a very flexible and powerful tool because of the scores, and therefore the final ranking is obtained by the pairwise relative evaluations of both the criteria and the options provided by the user. The decision maker's experience always guides the computations made by the AHP, and the AHP can thus be considered as a tool that can translate the evaluations (both qualitative and quantitative) made by the decision maker into a multi-criteria ranking. Also, the

AHP is simple because there is no need for building a complex expert system with the decision maker's knowledge embedded in it.

On the other hand, the AHP may require a large number of evaluations by the user, especially for problems with many criteria and options. Although every single evaluation is very simple since it only requires the decision maker to express how two options or criteria are compared to each other, the load of the evaluation task may become unreasonable. The number of pairwise comparisons grows quadratically with the number of criteria and options. For instance, when comparing ten alternatives on four criteria, 4·3/2 = 6 comparisons are requested to build the weight vector, and 4·(10·9/2) = 180 pairwise comparisons are needed to build the score matrix.

However, in order to reduce the decision maker's workload, the AHP can be completely or partially automated by specifying suitable thresholds for automatically deciding some pairwise comparisons.

3.12.4 Implementation of the analytic hierarchy process

The AHP can be implemented in three simple consecutive steps:

1 Computing the vector of criteria weights.
2 Computing the matrix of option scores.
3 Ranking the options.

Each step will be described in detail in the following. It is assumed that m evaluation criteria are considered, and n options are to be evaluated. A useful technique for checking the reliability of the results will also be introduced.

3.12.4.1 Computing the Vector of Criteria Weights

In order to compute the weights for the different criteria, the AHP starts creating a pairwise comparison matrix A. The matrix A is an $m \times m$ real matrix, where m is the number of evaluation criteria considered. Each entry a_{jk} of matrix A represents[16] the importance of the jth criterion relative to the kth criterion. If $a_{jk} > 1$, then the jth criterion is more important than the kth criterion, while if $a_{jk} < 1$, then the jth criterion is less important than the kth criterion. If two criteria have the same importance, then the entry a_{jk} is 1. The entries a_{jk} and a_{kj} satisfy the following constraint:

$$a_{jk} \cdot a_{kj} = 1 \tag{3.51a}$$

Obviously, $a_{jj} = 1$ for all j. The relative importance between two criteria is measured according to a numerical scale from 1 to 9, as shown in Table 3.30, where it is assumed that the jth criterion is equally or more important than the kth criterion. The phrases in the "Interpretation" column of Table 3.30 areonly suggestive and may be used to translate the decision maker's qualitative evaluations of

Table 3.30 Relative scores

Value of a_{jk}	*Interpretation*
1	j and k are equally important
3	j is slightly more important than k
5	j is more important than k
7	j is strongly more important than k
9	j is absolutely more important than k

the relative importance between two criteria into numbers. It is also possible to assign intermediate values which do not correspond to a precise interpretation. The values in the matrix A are by construction pairwise consistent. On the other hand, the ratings may in general show slight inconsistencies. However, these do not cause serious difficulties for the AHP.

Once the matrix A is built, it is possible to derive from A the normalized pairwise comparison matrix A_{norm} by making equal to 1 the sum of the entries on each column, i.e. each entry $a_{jk}a$ of the matrix A_{norm} is computed as follows:

$$\bar{a}_{jk} = \frac{a_{jk}}{\sum_{i=1}^{m} a_{jk}} \tag{3.51b}$$

Finally, the criteria weight vector w (that is an m-dimensional column vector) is built by averaging the entries on each row of A_{norm}.

3.12.4.2 Computing the Matrix of Option Scores

The matrix of option scores is an $n \times m$ real matrix S. Each entry s_{ij} of S represents the score of the ith option concerning the jth criterion. In order to derive such scores, a pairwise comparison matrix B^j is first built for each of the m criteria, $j = 1,\ldots,m$. The matrix B^j is an $n \times n$ real matrix, where n is the number of options evaluated. Each entry b_{ih}^{j} of the matrix B_j represents the evaluation of the ith option compared to the hth option concerning the jth criterion. If $b_{ih}^{j} \succ 1$, then the ith option is better than the hth option, while if $b_{ih}^{j} \prec 1$, then the ith option is worse than the hth option. If two options are evaluated as equivalent concerning the jth criterion, then the entry b_{ih}^{j} is 1. The entries b_{ih}^{j} and b_{hi}^{j} satisfy the following constraint:

$$b_{ih}^{j} \cdot b_{hi}^{j} = 1 \tag{3.52}$$

Moreover, $b_{ii}^{j} = 1$ for all i. An evaluation scale similar to the one introduced in Table 3.1 may be used to translate the decision maker's pairwise evaluations into numbers.

Second, the AHP applies to each matrix B^j the same two-step procedure described for the pairwise comparison matrix A, i.e. it divides each entry by the sum of the entries in the same column, and then it averages the entries on each row, thus obtaining the score vectors $s^j, j = 1,\ldots,m$.

The vector s^j contains the scores of the evaluated options concerning the jth criterion.

Finally, the score matrix $S$is obtained.

$$s = \left[s^{(1)} \ldots s^{(m)} \right] \tag{3.53}$$

i.e. the jth column of S corresponds to $S^{(j)}$.

Remark. In the considered DSS structure, the pairwise option evaluations are performed by comparing the values of the performance indicators corresponding to the decision criteria. Hence, this step of the AHP can be considered as a transformation of the indicator matrix I into the score matrix S.

3.12.4.3 Ranking the Options

Once the weight vector w and the score matrix S have been computed, the AHP obtains a vector v of global scores by multiplying S and w, i.e.

$$v = S \cdot w \tag{3.54}$$

The ith entry vi of v represents the global score assigned by the AHP to the ith option. As the final step, the option ranking is accomplished by ordering the global scores in a decreasing order.

3.12.5 Checking the Consistency

When many pairwise comparisons are performed, some inconsistencies may typically arise. One example is the following. Assume that three criteria are considered, and the decision maker evaluates that the first criterion is slightly more important than the second criterion, while the second criterion is slightly more important than the third criterion. An evident inconsistency arises if the decision maker evaluates by mistake that the third criterion is equally or more important than the first criterion. On the other hand, a slight inconsistency arises if the decision maker evaluates that the first criterion is also slightly more important than the third criterion. A consistent evaluation would be, for instance, that the first criterion is more important than the third criterion.

The AHP incorporates an effective technique for checking the consistency of the evaluations made by the decision maker when building each of the pairwise comparison matrices involved in the process, namely the matrix A and the matrices B^j. The technique relies on the computation of a suitable consistency index (CI) and will be described only for the matrix A. It is straightforward to adapt it to the case of the matrices B^j by replacing A with B^j,w with $S^{(j)}$, and m with n.

The CI is obtained by first computing the scalar x as the average of the elements of the vector whose jth element is the ratio of the jth element of the vector $A.w$ to the corresponding element of the vector w. Then,

$$\text{CI} = \frac{x - m}{m - 1} \tag{3.55}$$

A perfectly consistent decision maker should always obtain CI = 0, but small values of inconsistency may be tolerated. In particular, if

$$\frac{\text{CI}}{\text{RI}} \prec 0.1 \tag{3.56}$$

The inconsistencies are tolerable, and a reliable result may be expected from the AHP. In (8), RI is the Random Index, i.e. the CI when the entries of A are completely random. The values of RI for small problems ($m \leq 10$) are shown in Table 3.31.

The matrices A corresponding to the cases considered in the aforementioned example are shown below, together with their consistency evaluation based on the computation of the CI. Note that the conclusions are as expected.

$$\begin{gathered}
A = \begin{pmatrix} 1 & 3 & 1/3 \\ 1/3 & 1 & 3 \\ 3 & 1/3 & 1 \end{pmatrix} \Rightarrow \text{CI/RI} = 1.150 \Rightarrow \text{Inconsistent} \\
A = \begin{pmatrix} 1 & 3 & 3 \\ 1/3 & 1 & 3 \\ 1/3 & 1/3 & 1 \end{pmatrix} \Rightarrow \text{CI/RI} = 0.118 \Rightarrow \text{Slightly Inconsistent} \\
A = \begin{pmatrix} 1 & 3 & 5 \\ 1/3 & 1 & 3 \\ 1/5 & 1/3 & 1 \end{pmatrix} \Rightarrow \text{CI/RI} = 0.033 \Rightarrow \text{Consistent}
\end{gathered} \tag{3.57}$$

3.12.6 Automating the Pairwise Comparisons

Although every single AHP evaluation is very simple (the decision maker is only required to express how two criteria or alternatives are compared to each other), the load of the evaluation task may become unreasonable and tedious for the decision maker when many criteria and alternatives are considered. However, in order to alleviate the decision maker's workload, some pairwise comparisons

Table 3.31 Values of the Random Index (RI) for small problems

M	*2*	*3*	*4*	*5*	*6*	*7*	*8*	*9*	*10*
RI	0	0.58	0.90	1.12	1.24	1.32	1.41	1.45	1.51

can be completely or partially automated. A simple method is suggested in the following:

Let the *j*th criterion be expressed by an attribute which assumes values in the interval $\left[I_{j,\min}, I_{j,\max}\right]$ and let $I_j^{(i)}$ and $I_j^{(h)}$ be the instances of the attribute under the *i*th and *h*th control options, respectively. Assume that the larger the value of the attribute, the better the system performance according to the *j*th criterion. If $I_j^{(i)} \geq I_j^{(h)}$, the element $b_{ih}^{(i)}$ of B^j can be computedas follows:

$$b_{ih}^{(j)} = \frac{I_j^{(i)} - I_j^{(h)}}{I_{j,\max} - I_{j,\min}} + 1 \tag{3.58}$$

A similar expression holds if the smaller the value of the attribute, the better the system performance according to the j_{th} criterion. If $I_j^{(i)} \leq b_j^{(h)}$, the element $b_{ih}^{(i)}$ of B^j can be computed as follows:

$$b_{ih}^{(j)} = \frac{I_j^{(h)} - I_j^{(i)}}{I_{j,\max} - I_{j,\min}} + 1 \tag{3.59}$$

Note that (10) and (11) are linear functions of the difference $I_{ij} - I_{hj}$. Of course, more sophisticated functions can be designed by exploiting specific knowledge and experience.

3.12.7 *An illustrative example*

An example will be here described in order to clarify the mechanism of the AHP. $m = 3$ evaluation criteria are considered, and $n = 3$ alternatives are evaluated. Each criterion is expressed by an attribute. The larger the value of the attribute, the better the performance of the option concerning the corresponding criterion. The decision maker first builds the following pairwise comparison matrix for the three criteria:

$$A = \begin{pmatrix} 1 & 3 & 5 \\ 1/3 & 1 & 3 \\ 1/5 & 1/3 & 1 \end{pmatrix} \tag{3.60}$$

to which corresponds the weight vector $w = [0.633\ 0.261\ 0.106]^T$. Then, based on the values assumed by the attributes for the three options (see Figure 3.9), the decision maker builds the following pairwise comparison matrices:

$$B^{(1)} = \begin{pmatrix} 1 & 3 & 7 \\ 1/3 & 1 & 5 \\ 1/7 & 1/5 & 1 \end{pmatrix},\quad B^{(2)} = \begin{pmatrix} 1 & 1/5 & 1 \\ 5 & 1 & 5 \\ 1 & 1/5 & 1 \end{pmatrix},\quad B^{(3)} = \begin{pmatrix} 1 & 5 & 9 \\ 1/5 & 1 & 3 \\ 1/9 & 1/3 & 1 \end{pmatrix} \tag{3.61}$$

to which correspond the score vectors $s^{(1)} = [0.643\ 0.283\ 0.074]^T$, $s^{(2)} = [0.143\ 0.714\ 0.143]^T$, and $s^{(3)} = [0.748\ 0.180\ 0.072]^T$.

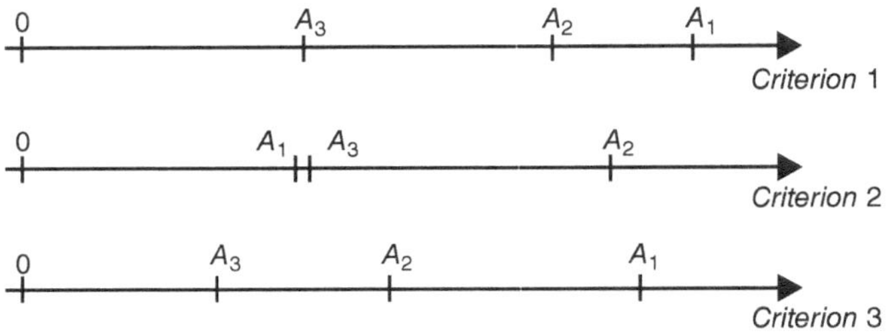

Figure 3.9 Values of the attributes for the alternatives A1, A2 and A3 (the scale on each axis is not relevant).

Hence, the score matrix *S* is

$$S = \begin{bmatrix} s^{(1)} & s^{(2)} & s^{(3)} \end{bmatrix} = \begin{pmatrix} 0.643 & 0.143 & 0.748 \\ 0.283 & 0.714 & 0.180 \\ 0.074 & 0.143 & 0.072 \end{pmatrix} \tag{3.62}$$

and the global score vector is $\boldsymbol{v} = \boldsymbol{S} \cdot \boldsymbol{w} = [0.523\ 0.385\ 0.092]^T$. Note that the first option turns out to be the most preferable, though it is the worst of the three concerning the second criterion.

References: Saaty, T.L. *The Analytic Hierarchy Process.* New York: McGraw-Hill, 1980.

DEMATEL.[17] This technique was developed in late 1971 mainly to examine complex global issues. Objectives of the strategic and objective of global issues were considered in order to access the appropriate solutions and used scientific, political, economic, social, ideological leaders, and artists for judgment and review. Interviews and questionnaires were repeatedly used to access expert judgment.

Three different types of questions were also used in the Server: questions about the features and indicators (or solutions) effective from a given problem, questions about the possible relationships between the indicators (or different issues), by specifying its severity Cardinal relationships (scoring), and questions to examine the nature of the identified elements and critique them for probability and revisit.

Decision Making Trial and Evaluation Laboratory (DEMATEL) is also used to construct a sequence of supposed information. So that the severity of communications is evaluated in the form of scoring, it tracks feedback with their significance and accepts irreversible relationships although empirical information has shown that the judgment of the experts of the direct communication of the elements together provides more or less transferable properties.

The DEMATEL method, developed by the Science and Human Affairs Program of the Battelle Memorial Institute of Geneva between 1972 and 1976, was used to research and solve complicated and intertwined problem groups. The applicability of the DEMATEL method is widespread, ranging from analyzing problematic world DM to industrial planning. The most important property of the DEMATEL method used in the multi-criteria decision making (MCDM) field is to construct interrelations between criteria.

After the interrelations between criteria were determined, the results derived from the DEMATEL method could be used for fuzzy integrals to measure the super-additive effectiveness value or for the Analytic Network Process (ANP) method to measure dependence and feedback relationships between certain criteria. There are four steps in the DEMATEL method:

1 calculate the average matrix;
2 calculate the normalized initial direct-influence matrix;
3 derive the total relation matrix;
4 set a threshold value and obtain the impact-relations map.

In Step 4, an appropriate threshold value is necessary to obtain a suitable impact-relations map as well as adequate information for further analysis and DM. The traditional method followed to set the threshold value is conducting discussions with experts. The results of the threshold values may differ among different researchers.

In contrast to the traditional method, which confronts the loop from a "set a threshold value" to obtain "the needed impact-relations-map," we propose the Maximum Mean De-Entropy (MMDE[18]) algorithm to obtain a unique threshold value for delineating the impact-relations map. In the numerical examples, real cases are used to discover and illustrate the intertwined effects of a structural multi-criteria decision making model.

The rest of this chapter is organized as follows. Section 2 briefly describes the DEMATEL method. The steps of the MMDE algorithm will be described, explained, and discussed in Section 3. In Section 4, numerical examples are shown in order to explain the proposed algorithm and discuss the results. Finally, in Section 5, we conclude.

The end product of the DEMATEL process, the impact-relations map, is a visual representation of the mind by which the respondent organizes his or her action in the world. This organizational process must occur for the respondent to keep internally coherent and to reach his or her personal goals. The steps of the DEMATEL method are described as follows:

Step 1. Find the average matrix. Suppose there are h experts available to solve a complex problem and there are n factors to be considered. The scores are given by each expert and give us an $n \times n$ nonnegative answer matrix X^k, with $1 \le k \le h$. Thus, X^1, X^2,...,X^h denote the answer matrices for each of the h experts, and each element of X^k is an integer denoted by $k_{ij} \cdot x$. The diagonal elements of each answer matrix X^k are all set to zero. We can then compute the $n \times n$ average matrix A by averaging the h experts score matrices. The (i, j) element of matrix A is denoted by a_{ij}:

$$a_{ij} = \frac{1}{h} \sum_{h=1}^{k} x_{ij}^{k} \tag{3.63}$$

In application, respondents were asked to indicate the direct-influence that they believe each factor exerts on each other according to an integer scale ranging from 0 to 4. A high score from a respondent indicates a belief that greater improvement in i is required to improve j. From any group of direct matrices of respondents, it is possible to derive an average matrix A.

Step 2. Calculate the normalized initial direct-relation matrix. We then create a matrix D by using a simple matrix operation on A. Suppose we create matrix D and $D = s \cdot A$, where

$$s = \text{Min}\left[\frac{1}{\max_{1\le i\le n}\sum_{j=1}^{n}\left|a_{ij}\right|}, \frac{1}{\max_{1\le j\le n}\sum_{i=1}^{n}\left|a_{ij}\right|}\right] \tag{3.64}$$

Matrix D is called the normalized initial direct-relation matrix. The (i, j) element d_{ij} denotes the direct-influence from factor x_i to factor x_j. Suppose d_i denotes the row sum of ith row of matrix D.

$$d_{i*} = \sum_{j=1}^{n} d_{ij} \tag{3.65}$$

d_i shows the sum of influence directly exerted from factor x_i to the other factors. Suppose d_j denotes the column sum of the jth column of matrix D.

$$d_{*j} = \sum_{i=1}^{n} d_{ij} \tag{3.66}$$

Then, d_j shows the sum of influence that factor x_j received from the other factors. We can normalize d_i and d_j as follows:

$$\begin{aligned} w_i(d) &= \frac{d_{i*}}{\sum_{i=1}^{n} d_{i*}} \\ v_j(d) &= \frac{d_{*j}}{\sum_{j=1}^{n} d_{*j}} \end{aligned} \tag{3.67}$$

Matrix D shows the initial influence which a factor exerts and receives from another. Each element of matrix D portrays a contextual relationship among the elements of the system and can be converted into a visible structural model of an impact-relations map of the system concerning that relationship. For example, as shown in Figure 3.10, the respondents are requested to indicate only direct links. In the directed graph represented in Figure 3.10, factor i directly affects only factors j and k, while indirectly, it also affects first l, m, and n and, second, o and q. The digraph map helps to explain the structure of the factors.

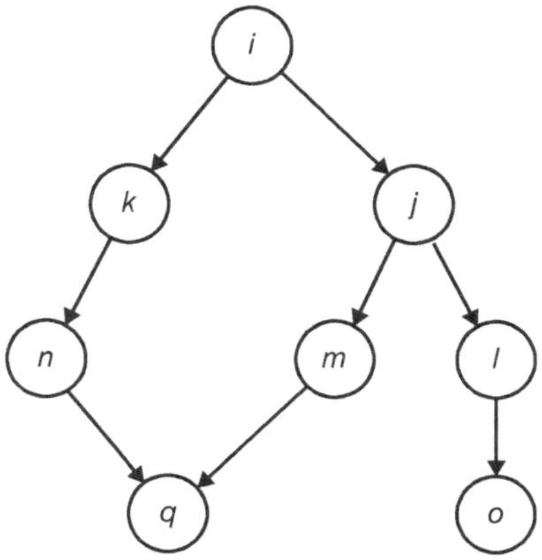

Figure 3.10 Example of a direct graph.

Step 3. A continuous decrease of the indirect effects of problems along the powers of matrix D, e.g. $D^2, D^3, \ldots, D^\infty$. It guarantees convergent solutions to the matrix inversion, similar to an absorbing Markov chain matrix. Note that $\lim_{m \to \infty} D^m = [0]_{n \times n}$ were $[0]_{n \times n}$ which is the $n \times n$ null matrix. The total relation matrix T is an $n \times n$ matrix and is defined as follows:

$$\begin{aligned}\sum_{m=1}^{\infty} D^i &= D + D^2 + D^3 + \cdots + D^m \\ &= D(I + D + D^2 + D^3 + \cdots + D^{m-1}) \\ &= D(I - D)^{-1}(I - D)(I + D + D^2 + D^3 + \cdots + D^{m-1}) \\ &= D(I - D)^{-1}(I - D^m) = D(I - D)^{-1}\end{aligned} \tag{3.68}$$

where I is the identity matrix, and T is called the total relation matrix. The (i, j) element of the matrix T, t_{ij}, denotes the full direct- and indirect-influence exerted from factor x_i to factor x_j.

Step 4. Set a threshold value and obtain the impact-relations map. In order to explain the structural relationship among the factors while keeping the complexity of the system to a manageable level, it is necessary to set a threshold value p to filter out the negligible effects in matrix T. Using the values of $w_i(t)$ and $v_i(t)$ from the matrix of full direct/indirect-influence relations, the level of dispatching and receiving of the influence of factor i can be defined. The interrelationship of each factor can be visualized as the oriented graphs on a two-dimensional plane after a certain threshold is set. Only those factors that affect matrix T greater than the threshold value should be chosen and shown in an impact-relations map.

In Step 4, the threshold value can be chosen by the decision maker or through discussions with experts. If the threshold value is too low, the map will be too complex to show the necessary information for DM. If the threshold value is too high, many factors will be presented as independent factors, without showing the relationships with other factors. Each time the threshold value increases, some factors or relationships will be removed from the map (an example based on a total relation matrix T is shown in Figure 3.11). An appropriate threshold value is necessary to obtain a suitable impact-relations map as well as adequate information for further analysis and DM.

$$T^{\text{example}} = \begin{bmatrix} 0.0093 & 0.0126 & 0.0538 & 0.0523 & 0.0759 \\ 0.0284 & 0.0077 & 0.0292 & 0.0284 & 0.0517 \\ 0.0509 & 0.0729 & 0.0087 & 0.0299 & 0.0341 \\ 0.0313 & 0.0340 & 0.0531 & 0.0086 & 0.0752 \\ 0.0532 & 0.0758 & 0.0547 & 0.0532 & 0.0150 \end{bmatrix} \tag{3.69}$$

Reference: Li, Chung-Wei, and Gwo-Hshiung Tzeng. Identification of a threshold value for the DEMATEL method using the maximum mean de-entropy algorithm to find critical services provided by a semiconductor intellectual property mall. *Expert Systems with Applications*,2009. Vol. 36, No.6, pp. 9891–9898.

NAIADE.[19] The name of this method of combining the beginning of words with the term "New Perspective" was derived from inaccurate estimation and DM environments, and was developed by Munda in 1995 to solve multi-criteria problems. This method, by considering the appropriate criteria and through complex, variable, and fuzzy measurements, compares the alternatives and their valuation, and because of the aforementioned characteristics, it is a suitable method for solving environmental problems.

There is also a step in analyzing conflict. Since there are disagreements between experts on environmental issues, the ability to use a conflict analysis,

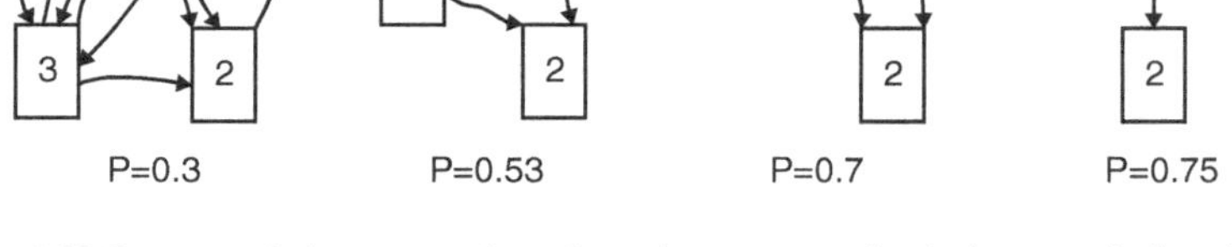

Figure 3.11 Impact-relations maps based on the same total relation matrix but different threshold values.

along with a multi-criteria assessment of proposed solutions, has identified Novel Approach to Imprecise Assessment and Decision Environments (NAIADE) as a powerful technique for solving environmental problems. Using this method, decision makers can make defensible decisions that will create the least conflict and increase the degree of satisfaction of different social groups.

NAIADE is a discrete multi-criteria method developed by G. Munda (1995) whose impact (or evaluation) matrix may include either crisp, stochastic, or fuzzy measurements of the performance of an alternative with respect to an evaluation criterion; thus, it is very flexible for real-world applications.

A peculiarity of NAIADE is the use of conflict analysis procedures to be integrated with the multi-criteria results. NAIADE can give the following information:

Ranking of the alternatives according to the set of evaluation criteria (i.e. technical compromise solution/s);
indications of the distance of the positions of the various interest groups (i.e. possibilities of convergence of interests or coalition formations);
ranking of the alternatives according to actors' impacts or preferences (social compromise solution).

From a methodological point of view, two main issues are faced:

the problem of equivalence of the used procedures in order to standardize the various evaluations (of a mixed type) of the performance of alternatives according to different criteria;
the problem of comparison of fuzzy numbers typical of all fuzzy multi-criteria methods.

These two methodological issues are dealt with a new semantic distance that is useful in the case of continuous, convex membership functions also allowing a definite integration.

If $mA1(x)$ and $mA2(x)$ are two membership functions, we can write

$$f(x) = k1mA1(x) \text{ and } g(y) = k2mA2(x),$$

where $f(x)$ and $g(y)$ are two functions obtained by rescaling the ordinates of $mA1(x)$ and $mA2(x)$ through $k1$ and $k2$, such as the area of $f(x)$ and $g(y)$ is equal to 1.

The distance between all points of the membership functions is computed as follows:

$$S_d(f(x), g(y)) = \int_x \int_y |y - x| f(x)g(y)d_y d_x \tag{3.70}$$

Therefore, when the intersection is empty, their distance is equal to the distance between their expected values. When the intersection between two fuzzy sets is not empty, their distance is greater than the difference between the expected. In this case, one finds

$$S_d(f(x), g(y)) = \int_{-\infty}^{+\infty}\int_{x}^{+\infty}(y-x)f(x)g(y)d_y d_x + \int_{-\infty}^{+\infty}\int_{-\infty}^{x}(x-y)f(x)g(y)d_y d_x \tag{3.71}$$

From a theoretical point of view, the following conclusions can be drawn:

1 the absolute value metric is a particular case of this type of distance;
2 the comparison between a fuzzy number and a crisp number is equal to the difference between the expected value of the fuzzy number and the value of the crisp number considered;
3 using this semantic distance, the problem of the use of only one side of the membership functions, common to most of the traditional fuzzy multi-criteria methods, is overcome;
4 also, stochastic information can be taken into account.

The whole NAIADE procedure can be dividedinto four main steps:

1 pairwise comparison of alternatives according to each criterion;
2 aggregation of all criteria;
3 ranking of alternatives;
4 social conflict analysis.

The comparison between the criterion scores of each pair of actions is carried out using the semantic distance. Another important advantage of the preference modeling introduced in NAIADE is toward sensitivity analysis. The modeling procedure based on the notion of a pseudo-criterion may present a serious lack of stability. Such undesirable discontinuities make a sensitivity analysis (or robustness analysis) necessary; however, this important analysis step is quite complex to manage because of the combinatorial nature of the various sets of data. One should combine variations of two thresholds and k scores of the M criteria. The use of fuzzy set approaches since small variations of input data (scores, thresholds) will modify in a continuous way the resulting preference model can allow one to avoid these drawbacks.

A first possible approach is to associate a fuzzy outranking relation with a pseudo-criterion as in ELECTRE III or to use a generalized criterion as in PROMETHEE. It has to be noted that both methods require that some parameters have to be fixed. A second approach may be the use of fuzzy criterion scores; however, in this case, the problem of comparison of fuzzy numbers arises. This is the reason why in NAIADE, a method based on both the underlying philosophy of a "fuzzy pseudo-criterion" and the use of fuzzy criterion scores is used (see Figures 3.12 and 3.13).

Given the information on the pairwise performance of the alternatives according to every single criterion, it is necessary to aggregate these evaluations in order to take all criteria into account simultaneously. Thisis done by using a kind of concordance index aggregating the various credibility degrees obtained according to the criteria used.

The final ranking of the alternatives in a complete or partial preorder (γ problem formulation) is obtained using the basic idea of positive (leaving) and negative (entering) flows of the PROMETHEE methods.

Finally, in the framework of equity analysis, it is possible to evaluate the impact of different actions on different income/interest groups (see Figures 3.14 and 3.15).

Reference: Munda, G. *Multi criteria evaluation in a fuzzy environment, Contributions to economics Series.* Heidelberg: Physica-Verlag, 1995.

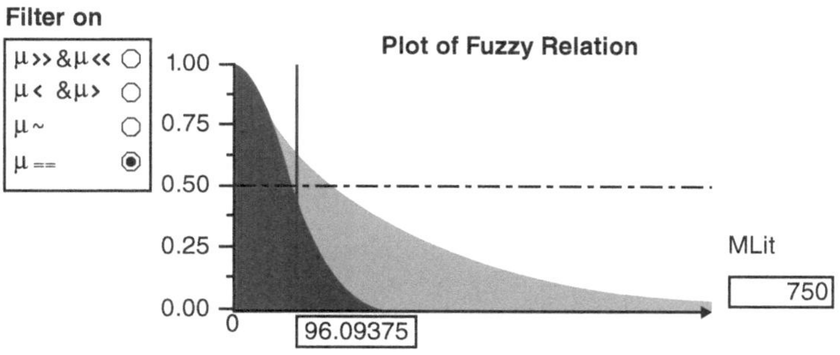

Figure 3.12 Credibility degrees of indifference relations.

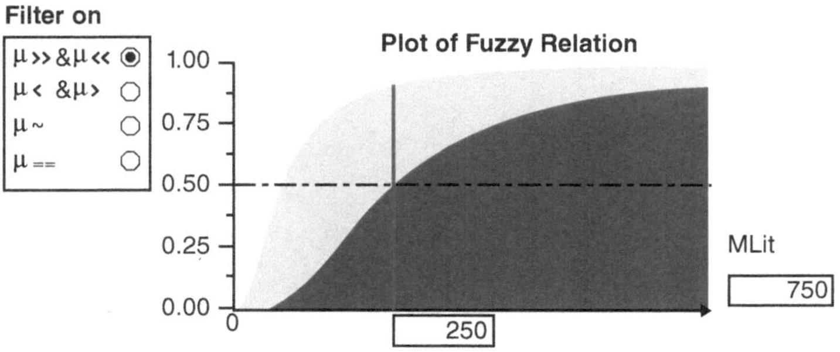

Figure 3.13 Credibility degrees of preference relations.

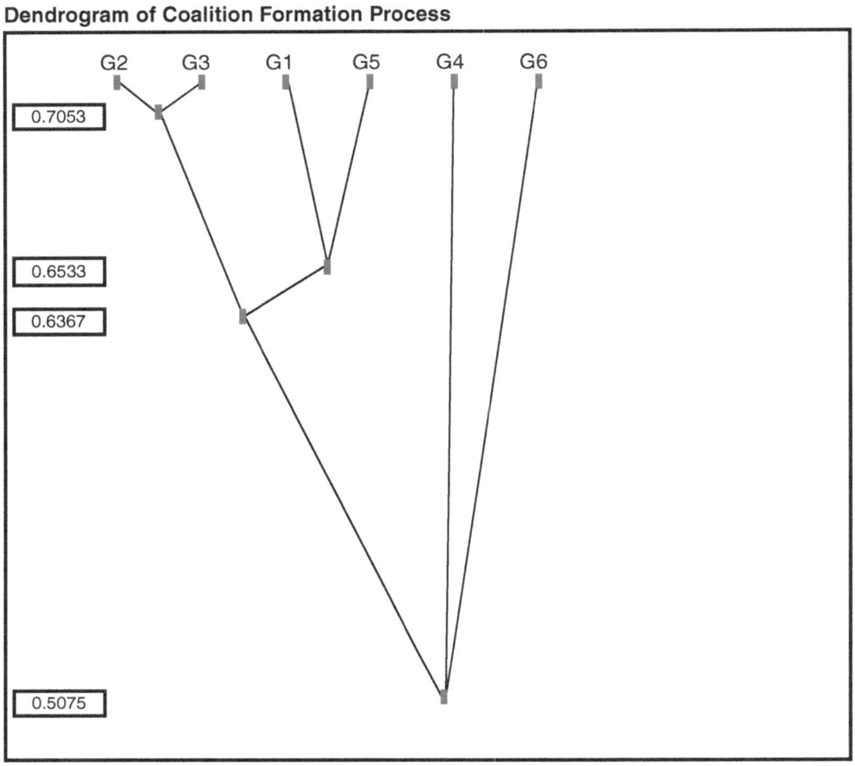

Figure 3.14 Credibility degrees of possible coalitions.

{ G3,G2,G5,G1 }	{ G4 }	{ G6 }
F 0.13	B 0.08	B 0.08
G 0.54	C 0.08	C 0.08
E 0.55	D 0.08	D 0.08
A 0.95	G 0.08	E 0.08
C 1.11	F 0.08	F 0.08
B 1.11	A 0.08	A 0.08
D 1.27	E 0.60	G 0.75

Figure 3.15 Ranking according to different possible coalitions.

3.13 Advanced Interactive Techniques

These techniques are advanced techniques designed to solve specific problems at a particular time. The various types of techniques can be mentioned in the following ways:

EVAMIX.[20] Voogd developed this technique in 1983. In this way, the alternatives are ranked according to the highest index. Determine the best index by comparing the pair of alternatives for all criteria and creating the Ordinal and Cardinal matrices.

From a procedural point of view, EVAMIX method consists of the seven steps discussed in the next section. It commences by identifying unique pairs (criterion-to-criterion) of alternatives. The degree of pairwise dominance for each pair of alternatives is calculated, as the difference in score received by the higher performing alternative compared to the poorer performing alternative.

The weighted sum of the dominance scores is then assigned to each alternative.

Step I. First, a set of objectives is identified. Then, various attributes and alternatives are shortlisted for the given application. Using this information, construct a data matrix of $(m \times n)$ size, where n is the number of alternatives and m is the number of relative attributes chosen for the selection problem. Next step is to distinguish the ordinal and cardinal criteria out of decision matrix. Attributes are given the linguistic preference, and can be converted into its corresponding crisp number as suggested by Chen and Hwang.

Step II. Normalizing the data set is done in the range of 0–1 using a linear normalization procedure. Different equations weigh the beneficial and non-beneficial attributes. For beneficial attributes, normalize the decision matrix using the following equation:

- For beneficial attributes, normalize the decision matrix using the following equation:

$$\frac{\left[x_{ij} - \min\left(x_{ij}\right)\right]}{\left[\max\left(x_{ij}\right) - \min\left(x_{ij}\right)\right]}; \quad (i = 1, 2, \ldots, m : j = 1, 2, \ldots, n) \tag{3.72}$$

- For non-beneficial attributes, the aforementioned equation can be rewritten as follows:

$$\frac{\left[\max\left(x_{ij}\right) - \left(x_{ij}\right)\right]}{\left[\max\left(x_{ij}\right) - \min\left(x_{ij}\right)\right]}; \quad (i = 1, 2, \ldots, m : j = 1, 2, \ldots, n) \tag{3.73}$$

According to this formulation in the normalized decision matrix, the maximum value will always be 1 and minimum value equal to 0.

Step III. Calculate the evaluative differences of *i*th alternative on each ordinal and cardinal criterion concerning other alternatives. This step involves the calculation of differences in criteria values between different alternatives pairwise. Pairwise is done based on the (AHP) by Saaty and Xu. It provides a way of breaking down the general data into a hierarchy of sub-data, which are easier to evaluate. These comparisons may be taken from actual measurements or from a fundamental scale which reflects the relative strength of preferences introduced by Fechner and further advocated by Turnstone (cited in Penwati, 1996).

In the pairwise comparison method, attributes and alternatives are presented in pairs. Evaluating individual alternatives is necessary. An attribute compared with it is always assigned the value 1, so the main diagonal entries of the pairwise comparison matrix are all 1. The numbers 3, 5, 7, and 9 correspond to the verbal judgments "moderate importance," "strong importance," "very strong importance," and "absolute importance" (with 2, 4, 6, and 8 for compromise between these values). The judgments are given using the fundamental scale of AHP.

- Let $A = [a_{ij}]$ for all I, j=1, 2,...,n ($a_i V_s a_j$) denote a square pairwise comparison matrix. Each entry in the matrix A is positive ($a_{ij} \succ 0$) and reciprocal $\left(a_{ij} = 1/a_{ji}, \forall_i, j = 1, 2, \ldots, n\right)$. Using the geometric mean method, weights are calculated by the following steps:
- Find the relative normalized weight (w_i) of each attribute by geometric means of rows in the matrix $A = [a_{ij}]$ and represent by A_1.
- Calculate matrices A_2 and A_3, where

$$A_2 = A \times A_1 \text{ and } A_3 = A_2 / A_1,$$
$$\text{where,} \tag{3.74}$$
$$A_1 = [w_1, w_2, \ldots, w_j]^T$$

- Determine the maximum Eigen value $\lambda_{\max}$ that is the average of matrix A_3.
- Calculate CI

$$\mathrm{CI} = (\lambda_{\max} - m) / (m - 1) \tag{3.75}$$

- The smaller the value of CI, the smaller is the deviation from the consistency.
- Obtain the RI for the number of attributes used in DM by T. L. Saaty (1980) and S. Xu (1986).
- Calculate the consistency ratio of CR = CI/RI. Usually, a CR of 0.1 or less is considered acceptable, and it reflects an informed judgment attributable to the knowledge of the analyst regarding the problem under study.

Step IV. Compute the dominance scores of each alternative pair, (i, i'), for all the ordinal and cardinal criteria using the following equations:

$$a_{ii'} = \left[\sum_{j \in o} \left\{ W_j \operatorname{sgn}\left(r_{ij} - r_{i'j}\right) \right\}^c \right]^{1/c}$$

$$\operatorname{sgn}\left(r_{ij} - r_{i'j}\right) = \begin{cases} +1 & \text{if} \quad r_{ij} \succ r_{i'j} \\ 0 & \text{if} \quad r_{ij} = r_{i'j} \\ -1 & \text{if} \quad r_{ij} \prec r_{i'j} \end{cases} \tag{3.76}$$

$$\gamma_{ii'} = \left[\sum_{j \in c} \left\{ W_j \operatorname{sgn}\left(r_{ij} - r_{i'j}\right) \right\}^c \right]^{1/c}$$

The symbol c denotes an arbitrary scaling parameter, for which any arbitrary positive odd number like 1, 3, 5... may be chosen, O and C are the sets of ordinal and cardinal criteria, respectively, and $a_{ii'}$ and $\gamma_{ii'}$ are the dominance scores for alternative pair, (i, i'), concerning ordinal and cardinal criteria, respectively. In order to be consistent, the same value of scaling parameter c is this formulation. It is assumed that the value of c for qualitative evaluation $a_{ii'}$ is equal to 1.

All standardized scores should have the same direction, *i.e.* a "higher" score should imply a "large" one.

Preference. It should be noted that the scores $\gamma_{ii'}$ of the quantitative criteria also have to represent "the higher, the better."

Step V. Since $a_{ii'}$ and $\gamma_{ii'}$ will have different measurement units, a standardization into the same unit is necessary. The standardized dominance scores can be writtenas follows:

$$\delta_{ii'} = h(a_{ii'}) \text{and}\, d_{ii'} = h(\gamma_{ii'}) \tag{3.77a}$$

where h represents a standardization function. The standardized dominance scores can be obtained using three different approaches, i.e. (a) subtractive summation technique, (b) subtracted shifted interval technique, and (c) additive interval technique. The following equations calculate the standardized ordinal score and cardinal dominance score for the alternative pair, (ii') using additive interval technique. Standardized ordinal dominance score:

$$(\delta_{ii'}) = \left[\frac{a_{ii'} - \bar{a}}{a^{+} - a^{-}}\right] \tag{3.77b}$$

where $a^{+}(a^{-})$ is the highest (lowest) ordinal dominance score for the alternative pair (ii')? Standardized cardinal dominance score:

$$(d_{ii'}) = \left[\frac{\gamma_{ii'} - \bar{\gamma}}{\gamma^{+} - \gamma^{-}}\right] \tag{3.78}$$

where $\gamma^{+}(\gamma^{-})$ is the highest (lowest) cardinal dominance score for the alternative pair (ii')?

Step VI. Let us assume that weight $j\, w$ has quantitative properties. The overall dominance measure ii D_{ii}' for each pair of alternatives (ii') is

$$D_{ii'} = w_{0}^{\delta_{ii'}} + w_{c} d_{ii'} \tag{3.79}$$

Where $O\, w$ is the sum of the weights for the ordinal criteria $D_{ii'} = \left(w_o = \sum_{j \in o} w_j \right)$

and w_c is the sum of the weights for the cardinal criteria $w_c = \left(w_o = \sum_{j \in c} w_j \right)$.

This overall dominance score reflects the degree to which alternative a_i dominates alternative a_i' for the given set of attributes and the weights. In general, the measure D_{ii}' may be considered as function K of the constituent appraisal scores.

This expression represents a well-known Pairwise comparison problem. Here for each pair, $D_{ii'} + D_{i'i} = 1$.

Step VII. Calculate the appraisal score. The appraisal score for the *i*th alternative (S_i) is computed which gives the final preference of the candidate alternatives. The higher the appraisal score, the better the performance of the alternatives. The best alternative is the one which has the highest value of the appraisal score.

$$(S_i) = \sum_{i'} \left(\frac{D_{i'i}}{D_{ii'}} \right)^{-1} \tag{3.80}$$

The methodology proposed in this chapter enables the decision maker to rank the alternatives from the best to the worst. The method can deal with any number of attributes and alternatives by effective mathematical steps.

MAVT.[21] This technique was developed by Keeny and Raiffa in 1996. In this method, alternatives are ranked using the value function $V(a_i)$. The function $V(a_i)$ is obtained by determining the value function for each criterion and calculating the value of the same criterion for a particular alternative.

Multiple attribute value theory (MAVT) can be used to address problems that involve a finite and discrete set of alternative policies that have to be evaluatedby conflicting objectives. For any given objective, one or more different attributes or criteria are used to measure the performance about that objective. These aspects, the impacts of all alternative options for all attributes, are presented in a so-called evaluation table. These attributes are usually measured on different measurement scales.

A closely related theory to MAVT is multiple attribute utility theory (MAUT). MAUT is based upon expected utility theory (von Winterfeldt and Edwards, 1986; French, 1988) and requires stronger assumptions to ensure additivity. The advantage of MAUT is that it can take uncertainty into account and represent it directly into its decision support model. MAUT is calleda strong form of DM and MAVT a weak form. However, MAUT is very difficult to apply, and no real applications are known. Therefore, we only focus here on MAVT.

MAVT is a compensatory technique. This means that the method does allow compensation of the weak performance of one criterion by a good performance of another criterion. MAVT aggregates the options' performance across all the criteria to form an overall assessment.

The intention of MAVT is to construct a means of associating a real number with each alternative, in order to produce a preference order on the alternatives consistent with decision maker value judgments. To do this, MAVT assumes that in every decision problem, a real value function U exists that represents the preferences of the decision maker. This function U is used to transform the attributes of each alternative policy into one single value. The alternative with the best value is pointed out as the best.

The process to be followed to carry out MAVT consists of the following four steps:

1 *Definition of alternatives.* Identify the policy alternatives which are to be compared with each other.

2 *Selection and definition of criteria.* Identify the effects or indicators relevant to the decision.
3 *Assessment of scores for each alternative in terms of each criterion.* Assign values to each effect or indicator for all alternatives.
4 *Ranking of the alternative.* A total score is calculated for each alternative by applying a value function *U* to all criteria's scores.

The first three steps are the same as in most MCA methods.[22] Step 4 is specific for MAVT.

3.13.1 Evaluation of results

MAVT is theoretically well established, and the simple additive form can be easily explained and is easy to use. The more complex forms of MAVT, however, are difficult to carry out, and experts are needed to apply the methods. MAVT is suitable to be applied in participatory processes.

The simple additive form of MAVT has been used to support the evaluation of a large number of problems all over the world (see Weighted Summation). Applications of the more complex form of MAVT are rare, probably due to the complexity of composing a value function that matches the problem at hand.

3.13.2 Policy Processes

MAVT is a useful tool to support the selection of a policy option. In its easier versions, it may be used in a relatively cheap and quick way. Its main advantage is that it is well axiomatized, and it can use the expertise accumulated in the Cost–Benefit Analysis, whose evaluation is also basedon the utility functions and welfare economics.

3.13.3 Sustainable Development Aspects

MAVT can be used to assess the sustainability of policy because it allows to simultaneously take into account indicators that refer to the three dimensions of sustainability, the economic, the social, and the economic one. In other words, it can be used to combine information in such a way that it can clarify sustainable development aspects. However, as it is a compensatory method, MAVT entails the evaluation process under weak sustainability assumptions.

MAVT can incorporate the following SD aspects as separate criteria to compare alternative policies: (de-) coupling aspects, adaptability, and (ir-) reversibility. MAVT is suitable to compare impacts independently of the gauge year. MAVT can incorporate the impacts on different groups/sectors/regions as separate categories and give a clear overview of the differences for these sectors. MAVT is suitable to compare impacts independently of the spatial dimension, as long as the spatial dimensions of the separate criteria are comparable. Moreover,-finally, the simplified additive MAVT can be applied to spatial data (Herwijnen, 1999; Sharifi and Herwijnen, 2002).

3.13.4 Operational Aspects

It is very difficult, if not impossible, to provide an accurate estimate of the costs of applying MAVT. However, as a very rough approximation, these can be expected now to mount about 15,000 euros. Thisis measured without gathering data. The costs of gathering data totally depend on the type of problem and cannot be estimated here. About two months are needed to apply MAVT on the data gathered. One week to structure the problem and the remaining time to compose a value function that can be used to transform the data. Because ideally gathering data can only be done after the problem is structured, the time needed for an assessment is much longer and totally depends on the time needed to gather the data (estimated between 1 and 12 months).

Furthermore, a large amount of data is needed to estimate the impacts. If quantitative data are not available, expert judgment can be used to estimate the impacts on a qualitative scale. MAVT can handle quantitative as well as qualitative data.

The main results of MAVT are a rank order of policy options and a better understanding of the problem at hand. The complex format of MAVT is not easy to grasp, and therefore the transparency of the tool is limited. Experts are needed to support the process of structuring the problem and to compose the value function. Several computer programs for MAVT intend to increase the user-friendliness of the tools as well as its reliability. Most of these software programs can also support the analysis of the uncertainty of the scores, weights, and value function(s).

The marginality of the effects of different policy options can be identified because MAVT compares each policy alternative with other alternatives and the current situation and assesses to which extent such effects are significant.

Because the criteria are defined by the user and depend on the problem, MAVT can be used in different countries, geographic levels, and levels of aggregation. One condition, however, is that the criteria should be definedin comparable countries, geographic levels, and levels of aggregation. MAVT has no mandatory use.

The time the results become outdated depends totally on the input data and the context of the application, so this cannot be established. MAVT looks at impacts of one selected year, which year is up to the user.

Because the user defines the criteria and depends on the problem, MAVT is independent of country, geographic level, and level of aggregation. One condition, however, is that the criteria should be definedin comparable countries, geographic levels, and levels of aggregation.

3.13.5 Experiences

The Keeney and Raiffa approach of the MAVT to decision support has been applied to many real decisions, in both the private and public sectors (Munda, 2005). Although well-regarded and effective, the nonadditive form is relatively complex and best implemented by specialists on major projects where time and expertise are both necessary and available.

What makes the Keeney and Raiffa model potentially demanding to apply is first that it takes uncertainty formally into account, building it directly into the decision support models and second that it allows attributes to interact with each other in other than a simple, additive fashion. It does not assume mutual independence of preferences. In this way, the analysis becomes more realistic (it represents a better way the human DM works), but also more complicated, so that in most cases, MAVT is used in its additive form, in order to allow a simpler and more transparent decision support, which can be implemented more quickly, by a wider range of users and for a larger set of problem types (Keeney and Raiffa, 1976). The simplified additive form of MAVT has been appliedmany times (Beinat and Nijkamp, 1998; ODPM, 2000[23]).

3.13.6 Combinations

MAVT, in general, can be included in the same packages of tools as Weighted Summation and the simplified additive form of MAVT. Experiences on this can be found there. However, experiences with the more complex form of MAVT as part of a package of tools are rare.

MAVT can be used to support the evaluation of alternative policies/plans/projects in policy impact assessment and Strategic Environmental Assessment.

3.13.7 Strengths and Weaknesses

3.13.7.1 Strengths

- MAVT helps in structuring a problem. By classifying the problem in various objectives, criteria to measure the objectives, and alternative options to solve the problem, MAVT provides a structured approach for dealing with the problem. MAVT also accommodates various types of information, quantitative as well as qualitative.
- MAVT enhances the understanding of the policy problem by forcing the decision maker and decision analyst to compose a value function that represents their preferences. This means that MAVT reduces the amount of information in order to improve its comprehensibility.
- Furthermore, MAVT provides a means of communication for reasoning and negotiations by clarifying the strengths and weaknesses of the alternative policies and by the possibility to visualize and communicate the intermediate and final results.
- Moreover, MAVT can incorporate the diverse views of stakeholder groups to construct the criteria tree, to develop alternative options/solutions for the problem, and to compose the value function.
- To apply an effective MAVT, good supporting software is essential (Belton and Stewart, 2002). Much software is available to apply the additive form of MAVT, weighted summation (see, for example, Janssen et al., 2001). An advantage of using software is the ease to test the robustness of the results using sensitivity analyses.

3.13.7.2 Weaknesses

- The main weakness of MAVT is that it assumes full compensability of criteria, that is, the criteria are all reduced and expressed in the same unit (in this case through value functions). This implies that bad performance on a criterion (for example, a high environmental impact) can be compensated by a good performance on another (for example, a high income). This feature makes MAVT an adequate instrument to operationalize the concept of weak sustainability, but not the idea of strong sustainability.
- The composition of the nonadditive value function in MAVT is a very difficult task, especially if the number of criteria involved is large and they are very different in character. A decision analyst is needed to support the decision makers to carry out this task.
- The mutual independence preference condition is often difficult to meet and, in many cases, simply ignored when applying the additive form of MAVT.

3.13.7.3 Further work

Applications of the nonadditive form of MAVT are not found. Examples within the EU context are also not found yet. These are necessary to demonstrate the use for the EU. As it holds for every MCA, MAVT compares different alternative options. Therefore, different alternative policy options are needed in an EU context, and that is often not the case. This aspect has to be worked out further. A first step has been made in the impact assessment guidelines developed by the EU (SEC, 2005) whom advise to consider alternative policy options instead of only one.

UTA.[24] Jacquet, Lagreze and Siskos developed this technique in 1982. This method is, in fact, a special case of the MAVT method and is used when the value function of each criterion is obtained using the Ordinal regression method.

The UTA (UTilités Additives) method proposed by Jacquet-Lagrèze and Siskos (1982) aims at inferring one or more additive value functions from a given ranking on a reference set A_R. The method uses special linear programming techniques to assess these functions so that the ranking(s) obtained through these functions on A_R is (are) as consistent as possible with the given one.

The criteria aggregation model in UTA is assumed to be an additive value function of the following form (Jacquet-Lagrèze and Siskos, 1982):

$$u(g)=\sum_{i=1}^{n} p_i u_i(g_i) \tag{3.81}$$

subject to normalization constraints.

$$\begin{cases} \sum_{i=1}^{n} p_i = 1 \\ u_i\left(g_{i^*}\right)=0, \quad u_i\left(g_i^*\right)=1 \quad \forall i=1,2,\ldots,n \end{cases} \tag{3.82}$$

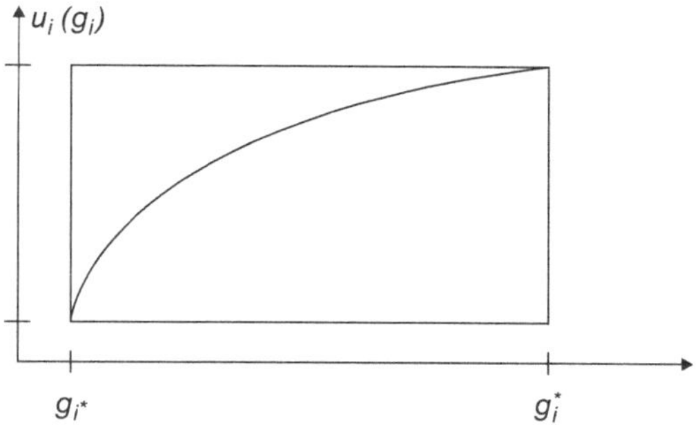

Figure 3.16 The normalized marginal value function.

where u_i, $i = 1,2,\ldots,n$ are non-decreasing real-valued functions, named marginal value or utility functions, which are normalized between 0 and 1, and p_i is the weight of u_i (Figure 3.16).

Both the marginal and global value functions have the monotonicity property of the true criterion. For instance, in the case of the global value function, the following properties hold:

$$\begin{cases} u[g(a)] \succ u[g(b)] \Leftrightarrow a \succ b(\text{perference}) \\ u[g(a)] = u[g(b)] \Leftrightarrow a = b(\text{perference}) \end{cases} \tag{3.83}$$

The UTA method infers an unweighted form of the additive value function, equivalent to the form defined from relations, as follows:

$$u(g) = \sum_{i=1}^{n} u_i(g_i) \tag{3.84}$$

subject to normalization constraints.

$$\begin{cases} \sum_{i=1}^{n} u_i\left(g_i^*\right) = 1 \\ u_i\left(g_{i^*}\right) = 0 \quad \forall_i = 1,2,\ldots,n \end{cases} \tag{3.85}$$

Of course, the existence of such a preference model assumes the preferential independence of the criteria for the DM (Keeney and Raiffa, 1976), while other conditions for additivity have been proposed by Fishburn (1966, 1967). This assumption does not pose significant problems in a posteriori analyses such as disaggregation analyses.

3.13.8 Development of the UTA Method

On the basis of the additive model and taking into account the preference conditions, the value of each alternative $a \in A_R$ may be written as follows:

$$u'[g(a)] = \sum_{i=1}^{n} u_i[g_i(a)] + \sigma(a) \quad \forall_a \in A_R \tag{3.86}$$

where $\sigma(a)$ is a potential error relative to $u'[g(a)]$.

Moreover, in order to estimate the corresponding marginal value functions in a piecewise linear form, Jacquet-Lagrèze and Siskos (1982) propose the use of linear interpolation. For each criterion, the interval $\left[g_{i^*}, g_i^*\right]$ is cut into $(a_i - 1)$ equal intervals, and thus the endpoints g_i^j are given by the following formula:

$$g_i^j = g_{i^*} + \frac{j-1}{a_i - 1}\left(g_i^* - g_{i^*}\right) \quad \forall_j = 1, 2, \ldots, a_i \tag{3.87}$$

The marginal value of an action is approximated by linear interpolation, and thus, for $g_i(a) \in \left[g_i^j, g_i^{j+1}\right]$

$$u_i[g_i(a)] = u_i\left(g_i^j\right) + \frac{g_i(a) - g_i^j}{g_i^{j+1} - g_i^j}\left[u_i\left(g_i^{j+1}\right) - u_i\left(g_i^j\right)\right] \tag{3.88}$$

The set of reference actions $A_R = \{a_1, a_2, \ldots, a_m\}$ is also "rearranged" in such a way that a_1 is the head of the ranking and a_m its tail. Since the ranking has the form of a weak order *R*, for each pair of consecutive actions (a_k, a_{k+1}), it holds either $a_k \succ a_{k+1}$ (preference) or $a_k \geq a_{k+1}$ (indifference).

Thus, if

$$\Delta(a_k, a_{k+1}) = u'\left[(g(a_k)\right] - u'\left[(g(a_{k+1})\right] \tag{3.89}$$

then one of the following holds:

$$\begin{cases} \Delta(a_k, a_{k+1}) \geq \delta & \text{if} \quad a_k \succ a_{k+1} \\ \Delta(a_k, a_{k+1}) = 0 & \text{if} \quad a_k = a_{k+1} \end{cases} \tag{3.90}$$

where δ is a small positive number so as to discriminate significantly two successive equivalence classes of *R*.

Taking into account the hypothesis on monotonicity of preferences, the marginal values $u_i(g_i)$ must satisfy the set of the following constraints:

$$u_i\left(g_i^{j+1}\right) - u_i\left(g_i^j\right) \geq s_i \quad \forall_j = 1, 2, \ldots, a_i - 1, i = 1, 2, \ldots, n \tag{3.91}$$

With $s_i \geq 0$ being indifference thresholds defined on each criterion g_i, Jacquet-Lagrèze and Siskos (1982) urge that it is not necessary to use these thresholds in the UTA model ($s_i = 0$), but they can be useful in order to avoid phenomena such as $u_i\left(g_i^{j+1}\right) = u_i\left(g_i^j\right)$ when $g_i^{j+1} \succ g_i^j$.

The marginal value functions are finally estimated by means of the following linear program (LP) with the spoken formulas as constraints and with an objective function depending on $\sigma(a)$ and indicating the amount of total deviation.

$$\begin{cases} [\min] F = \sum_{a \in A_R} \sigma(a) \\ \text{Subject to:} \\ \Delta(a_k, a_{k+1}) \geq \delta \quad \text{if} \quad a_k \succ a_{k+1} \\ \Delta(a_k, a_{k+1}) = 0 \quad \text{if} \quad a_k = a_{k+1} \\ u_i\left(g_i^{j+1}\right) - u_i\left(g_i^j\right) \geq 0 \quad \forall_i \text{ and } j \\ \sum_{i=1}^{n} u_i\left(g_i^*\right) = 1 \\ u_i\left(g_{i^*}\right) = 0, u_i\left(g_i^j\right) \geq 0, \forall_a \in A_R, \forall_i \text{ and } j \end{cases} \tag{3.92}$$

The stability analysis of the results provided by the above LPformula is considered as a post-optimality analysis problem. As Jacquet-Lagrèze and Siskos (1982) note, if the optimum $F^* = 0$, the polyhedron of admissible solutions for $u_i(g_i)$ is not empty, and many value functions lead to a perfect representation of the weak order R. Even when the optimal value F^* is strictly positive, other solutions, less good for F, can improve other satisfactory criteria, like Kendall's τ.

As shown in Figure 3.17, the post-optimal solutions space is defined by the polyhedron.

$$\begin{cases} F \leq F^* + k(F^*) \\ \text{All the constraints of LPin the previous formula} \end{cases} \tag{3.93}$$

where $k(F^*)$ is a positive threshold which is a small proportion of F^*.

The algorithms which could be used to explore the polyhedron in the aforementioned formula are branch and bound methods, like reverse simplex method (Van de Panne, 1975), or techniques dealing with the notion of the labyrinth in graph theory, such as Tarry's method (Charnes and Cooper, 1961), or the method of Manas and Nedoma (1968). Jacquet-Lagrèze and Siskos (1982), in the original UTA method, propose the partial exploration of a polyhedron by solving the following LPs:

$$\begin{cases} [\min] u_i\left(g_i^*\right) \\ \text{in} \\ \text{Polyhedron} \end{cases} \quad \begin{cases} [\max] u_i\left(g_i^*\right) \\ \text{in} \\ \text{Polyhedron} \end{cases} \quad \forall_i = 1, 2, \ldots, n \tag{3.94}$$

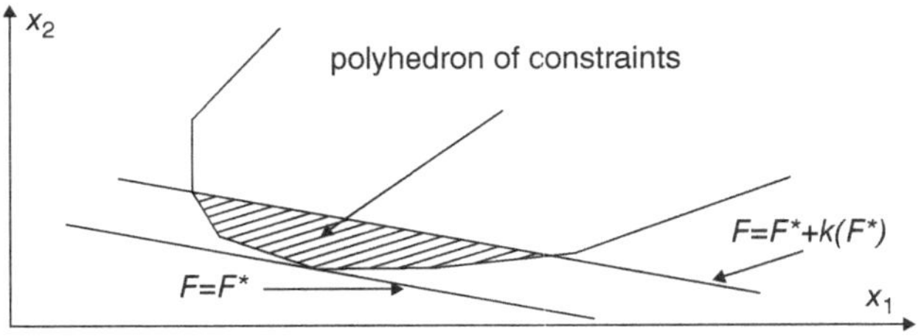

Figure 3.17 Post-optimality analysis (Jacquet-Lagrèze and Siskos, 1982).

The average of the previous LPs may be considered as the final solution to the problem. In case of instability, a large variation of the provided solutions appears, and this average solution is less representative. In any case, the solutions of the aforementioned LPs give the internal variation of the weight of all criteria g_i and consequently give an idea of the importance of these criteria in the DM's preference system.

3.13.9 The UTASTAR Algorithm

The UTASTAR method proposed by Siskos and Yannacopoulos (1985) is an improved version of the original UTA model presented in the previous section. In the original version of UTA (Jacquet-Lagrèze and Siskos, 1982), for each packed action $a \in A_R$, a single error $\sigma(a)$ is introduced to be minimized. This error function is not sufficient to completely minimize the dispersion of points all around the monotone curve in Figure 3.18. The problem is posed by points situated on the right of the curve, from which it would be suitable to subtract an amount of value/utility and not increase the values/utilities of the others.

In the UTASTAR method, Siskos and Yannacopoulos (1985) introduced a double positive error function, so that previous formulas become:

$$u'[g(a)] = \sum_{i=1}^{n} u_i [g_i(a)] - \sigma^+(a) + \sigma^-(a) \quad \forall_a \in A_R \tag{3.95}$$

where σ^+ and σ^- are the overestimation and the underestimation errors, respectively.

Moreover, another important modification concerns the monotonicity constraints of the criteria, which are taken into account through the transformations of the variables.

$$w_{ij} = u_i\left(g_i^{j+1}\right) - u_i\left(g_i^j\right) \geq 0 \quad \forall_i = 1, 2, \ldots, n \text{ and } j = 1, 2, \ldots, a_i - 1 \tag{3.96}$$

and thus, the previous monotonicity conditions can be replaced by the nonnegative constraints for the variables w_{ij} (for $s_i = 0$).

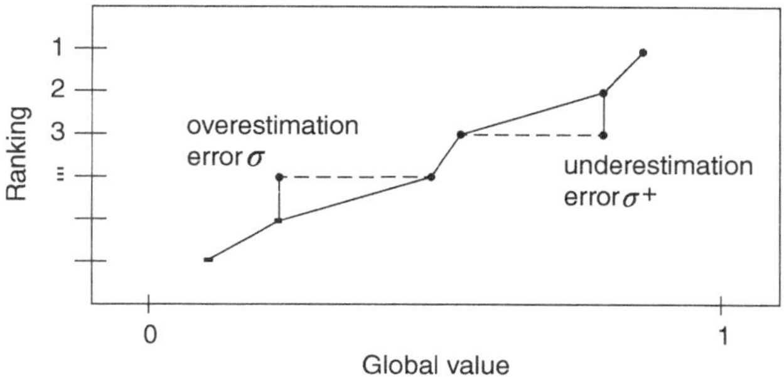

Figure 3.18 Ordinal regression curve (ranking versus global value).

Consequently, the UTASTAR algorithm may be summarized in the following steps:

Step 1.

Express the global value of reference actions $u\left[g\left(a_k\right)\right], \kappa = 1,2,\ldots,m$, first in terms of marginal values $u_i(g_i)$, and then in terms of variables w_{ij} according to the previous formula, employing the following expressions:

$$\begin{cases} u_i\left(g_i^1\right) = 0 & \forall_i = 1,2,\ldots,n \\ u_i\left(g_i^1\right) = \sum_{t=1}^{j-1} w_{it} & \forall_i = 1,2,\ldots,n \text{ and } j = 2,3,\ldots,a_i - 1 \end{cases} \tag{3.97}$$

Step 2.

Introduce two error functions σ^+ and σ^- on A_R by writing for each pair of consecutive actions in the ranking the analytic expressions.

$$\begin{aligned} \Delta\left(a_k, a_{k+1}\right) &= u\left[g\left(a_k\right)\right] - \sigma^+\left(a_k\right) + \sigma^-\left(a_k\right) \\ &- u\left[g\left(a_{k+1}\right)\right] + \sigma^+\left(a_{k+1}\right) - \sigma^-\left(a_{k+1}\right) \end{aligned} \tag{3.98}$$

Step 3.

Solve the LP.

$$\begin{cases} [\min] z = \sum_{k=1}^{m} \left[\sigma^+\left(a_k\right) + \sigma^-\left(a_k\right)\right] \\ \text{Subject to:} \\ \left.\begin{cases} \Delta\left(a_k, a_{k+1}\right) \geq \delta & \text{if} \quad a_k \succ a_{k+1} \\ \Delta\left(a_k, a_{k+1}\right) = 0 & \text{if} \quad a_k = a_{k+1} \end{cases}\right\} \Rightarrow \forall_k \\ \sum_{i=1}^{n} \sum_{j=1}^{a_i - 1} w_{ij} = 1 \\ w_{ij} \geq 0, \sigma^+\left(a_k\right) \geq 0, \sigma^-\left(a_k\right) \geq 0 \quad \forall_i, j \text{ and } k \end{cases} \tag{3.99}$$

with δ being a small positive number.

Step 4.

Test the existence of multiple or near-optimal solutions of the LP (stability analysis); in case of non-uniqueness, find the mean additive value function of those (near)-optimal solutions which maximize the objective functions.

$$u_i\left(g_i^*\right)=\sum_{j=1}^{a_i-1} w_{it} \quad \forall_i=1,2,\ldots,n \tag{3.100}$$

on the polyhedron of the constraints of the LP bounded by the new constraint.

$$\sum_{k=1}^{m}\left[\sigma^+(a_k)+\sigma^-(a_k)\right]\leq z^*+\varepsilon \tag{3.101}$$

where z^* is the optimal value of the LP in Step 3 and ε is a very small positive number.

A comparison analysis between UTA and UTASTAR algorithms is presented by Siskos and Yannacopoulos (1985) through a variety of experimental data. The UTASTAR method has provided better results concerning a number of comparison indicators, like:

1 The number of the necessary simplex iterations for arriving at the optimal solution.
2 The Kendall's τ between the initial weak order and the one produced by the estimated model.
3 The minimized criterion z (sum of errors) taken as the indicator of dispersion of the observations.

MAUT.[25] The MAUT method was developed by Keeny and Raiffa in 1976. This method is similar to the MAVT method. The only difference between these two methods is that in the MAUT technique, the values of the criteria are equal to their utility functions. The existence of this similarity forces decision makers to carefully consider their criteria.

The foundations of MAUT were laid by Churchman, Ackoff, and Arnoff (1957) who first treated a multiple-criteria decision problem using a SAW method. It was developed further using a quasi-additive and multi-linear utility function. Its early development is attributed to Fishburn (1968, 1970, 1978). Keeney and Raiffa devoted much of their work (e.g. 1976) to MAUT. Since then, the method has been developed further in the way that methodologies are elaborated to elicit the utility functions and to aggregate the single-criterion utility functions to a multi-attribute utility function, and now provides a formal mechanism for handling multiple objectives, intangible factors, risk, qualitative data, and time sequence effects in ex-ante appraisals based on the decision maker's preferences (Dillon and Perry, 1977). A frequently used software package, which is based upon the estimation of an additive utility function and on the interaction with the decision maker is the so-called PREFCALC software. Other software packages are 'Decide Right,' which is based on SMART, the Simple

Multi-Attribute Rating Technique (for explanation, see later in this chapter) and 'HIPRE 3+,'[26] which integrates two methods of decision analysis and problem-solving, namely AHP (see cross-references needed pp.) and SMART.[27]

As the name MAUT indicates, it is based on utility theory (von Neumann and Morgenstern 1947). It relies upon the basic von Neumann–Morgenstern axioms of preference and thus upon a utility function, which allows the comparison of risky outcomes through the computation of expected utility. Risk can arise in two ways, as risk about the outcomes or about the attribute values. MAUT can be used for the first form of risk1 (Hwang and Yoon 1981). MAUT uses directly assessed preferences with general aggregation; this involves direct questioning of the decision makers and the choice on the basis of an aggregate measure for each alternative (Dillon and Perry, 1977). It can be seen as an extension of ordinary subjective expected utility procedures to the case of a choice between multi-attributed alternatives.

To prepare a multiple attribute decision by the use of MAUT requires the following steps (Dillon and Perry, 1977):

1 specify the project alternatives (including combinations) as discrete entities;
2 elicit the decision makers set of probability distributions for outcomes associated with each project alternative in each attribute if there is a risk;
3 elicit the decision maker's utility function $u_i(x_i)$ for the range of outcomes on each attribute;
4 use the appropriate global multi-attribute utility function $U(x)$ to find the expected utility of each project alternative; and
5 choose the project or project combination with the highest expected utility; thus, the function U should be maximized.

Step 1. It comprises the definition of the alternatives, the objectives, the attributes, and the outcomes of each alternative in each attribute. This step is not specific to MAUT; this is how all discrete MCDA[28] methods start.

Step 2. It concerns risk viewed as the subjective probability. If there is risk concerning the impact of the actions on the attributes, probabilities p_{ij} can be assigned to the outcomes $x_1, y_1, \ldots$ of the alternatives on the attributes. It is assumed that the value of the action a for attribute i is given by the expected value of u_i (Vincke, 1985; Dillon and Perry, 1977).

$$u_i(x_i) = S_j p_{ij} u_i(x_{ij}) \tag{3.102}$$

where p_{ij} is the probability of the jth outcome in the ith attribute or in continuous terms, and

$$u_i(x_i) = \int u_i(x_i) f_i(x_i) dx_i \tag{3.103}$$

where $f_i(x_i)$ is the probability distribution of outcomes in the ith attribute.

There are different methods to elicit subjective probability distributions, which are not without difficulties. Which method to choose depends on the problem and often on the experience of the analysts. Direct involvement methods are simple and allow the decision makers to see and understand the built distribution. For these reasons, such direct involvement methods seem preferable to others, such as lottery-type questions (Dillon and Perry, 1977). The simplest method of all is asking for a few point estimates of probability; this is crude but practicable.

Step 3. It requires the elicitation of the decision maker's utility function $u_i(x_i)$ for the range of outcomes on each attribute. A utility function $u_i(x_i)$ is developed for each attribute. This is done by asking the decision maker a series of questions based on the axiom of continuity (Dillon and Perry, 1977, p. 10). Additionally, weights w_i are assigned to the attributes, presenting trade-offs, by asking for the decision makers order of importance of the attributes. The assessment of the appropriate utility function is complex and intricate. Thus, it is a crucial step in MAUT. This procedure leads us to three difficult questions (Vincke, 1985): (a) what must the properties of the decision maker's preferences be so that U is a certain function of u_i? (b) how to test the properties? and (c) how to construct the function U? The method can only be applied if these questions can be answered in a meaningful way.

After the establishment of the utility functions $u_i(x_i)$ and the elicitation of the attribute probability distributions, the expected utility is aggregated across the attribute distributions for each alternative.

Step 4. For each attribute, a function u_i is built and the functions u_i are aggregated in a global criterion U, such that $U\left(u_1\left(x_1\right),\ldots,u_n\left(x_n\right)\right) \succ \left(u_1\left(y_1\right),\ldots,u_n\left(y_n\right)\right)$ if the action represented by $(x_1, x_2,\ldots,x_n)$ is better than the action represented by $(y_1,y_2,\ldots,y_n)$, when considering all the attributes simultaneously (Vincke, 1985; Roy and Bouyssou, 1985). In that way, the multiple-criteria problem is reduced to a single-criterion decision problem. In order to build the aggregate function, different aggregation procedures are possible; the additive or the multiplicative aggregations are most widely applied. The additive form is the simplest form of a utility function, requiring two strong assumptions, namely utility independence and preferential independence of any subset of criteria (Fandel and Spronk, 1985).

SMART presents one method to build the additive form (Martel, in Climaco, 2005). If the additive model holds, the expected utility of each attribute for each alternative is added. Each attribute must get a weighting factor w_i to get a common scale for the utility functions $u_i(x_i)$. The utility for a certain alternative with uncertain consequences is then given by $U(x) = \sum w_i\left[u_i\left(x_i\right)\right]$ where in discrete terms, $u_i\left(x_i\right) = S_j p_{ij} u_i\left(x_{ij}\right)$ p_{ij} being the probability of the jth outcome in the ith attribute; or in continuous terms, $u_i\left(x_i\right) = \int u_i\left(x_i\right) f_i\left(x_i\right) dx_i$, where $f_i\left(x_i\right)$ is the probability distribution of outcomes in the ith attribute. If independence does not hold, the multiplicative form is usually applied. Another form is the quasi-additive model, which can be used if neither marginality nor preferential independence prevails, but this procedure becomes very complicated for cases with more than three attributes (Martel, in Climaco, 2005).

Step 5. It consists of summarizing the results and interpreting them.

SMART.[29] Oslon first used this method in 1996. This method is the applied mode of the MAUT method, and it is used when the optimal functions of each alternative are obtained by calculating the linear–linear mean of the decision makers' opinions.

The SMART technique is based on a linear additive model. This means that the overall value of a given alternative is calculated as the total sum of the performance score (value) of each criterion (attribute) multiplied with the weight of that criterion. The main stages in the analysis are as follows (adapted from Olson (1996)):

- ***Stage 1.*** Identify the decision maker(s).
- ***Stage 2.*** Identify the issue of issues. Utility depends on the context and purpose of the decision.
- ***Stage 3.*** Identify the alternatives. This step would identify the outcomes of possible actions, a data gathering process.
- ***Stage 4.*** Identify the criteria. It is important to limit the dimensions of value. This can be accomplished by restating and combining criteria, or by omitting less important criteria. It has been argued that it was not necessary to have a complete list of criteria. Fifteen was considered too many, and eight was considered sufficiently large. If the weight for a particular criterion is quite low, that criterion need not be included. There is no precise range of the number of criteria appropriate for decisions.
- ***Stage 5.*** Assign values for each criterion. For decisions made by one person, this step is fairly straightforward. The ranking is a decision task that is easier than developing weights, for instance. This task is usually more difficult in group environments. However, groups including diverse opinions can result in a more thorough analysis of relative importance, as all sides of the issue are more likely to be voiced. An initial discussion could provide all group members with a common information base. This could be followed by the identification of individual judgments of relative ranking.
- ***Stage 6.*** Determine the weight of each of the criteria. The most important dimension would be assigned an importance of 100. The next most important dimension is assigned a number reflecting the ratio of relative importance to the most important dimension. This process is continued, checking implied ratios as each new judgment is made. Since this requires a growing number of comparisons, there is a very practical need to limit the number of dimensions (objectives). It is expected that different individuals in the group would have different relative ratings.
- ***Stage 7.*** Calculate a weighted average of the values assigned to each alternative. This step allows normalization of the relative importance into weights summing to 1.
- ***Stage 8.*** Make a provisional decision.
- ***Stage 9.*** Perform sensitivity analysis.

In SMART, ratings of alternatives are assigned directly, in the natural scales of the criteria. For instance, when assessing the criterion “cost” for the choice between different road layouts, a natural scale would be a range between the most expensive and the cheapest road layout. In order to keep the weighting of the criteria and the rating of the alternatives as separate as possible, the different scales of criteria need to be converted into a common internal scale. In SMART, this is done mathematically by the decision maker by means of a value function. The simplest and most widely used form of a value function method is the additive model, which in the simplest cases can be applied using a linear scale (e.g. going from 0 to 100).

3.13.10 SMART Exploiting Ranks (SMARTER)

The assessment of value functions and swing weights in SMART can sometimes be a difficult task, and decision makers may not always be confident about it. Because of this, Edwards and Barron have suggested a simplified form of SMART named SMARTER (SMART Exploiting Ranks). Using the SMARTER technique, the decision makers place the criteria into an importance order. For example, ‘Criterion 1 is more important than Criterion 2, which is more important than Criterion 3, which is more important Criterion 4’ and so on, $C_1 \geq C_2 \geq C_3 \geq C_4 \ldots$ SMARTER then assign surrogate weights according to the Rank Order Distribution (ROD) method or one of the similar methods which are described below.

Barron and Barret (1996) believe that generated weights may be more precise than weights produced by the decision makers who may be more comfortable and confident with a simple ranking of the importance of each criterion swing, especially if it represents the considered outcome of a group of decision makers. Therefore, a number of methods that enable the ranking to be translated into ‘surrogate’ weights representing an approximation of the ‘true’ weights have been developed. A few of these methods are described below. Here, $w_j \succ 0$ are weights reflecting the relative importance of the ranges of the criteria values, where $\Sigma_j^n = 1 w_j = 1, i = 1, \ldots, n$ is the rank of the criteria, and n is the number of criteria in the decision problem.

Rank order centroid (ROC) weights. The ROC weights are defined by (Roberts and Goodwin, 2002)

$$w_j(\text{ROC}) = 1/2 \sum_{j=1}^{n} 1/j, i = 1, \ldots, n \tag{3.104}$$

Rank sum (RS) weights. The RS weights are the individual ranks normalized by dividing by the sum of the ranks. The RS weights are defined by (Ibid.)

$$w_j(\text{RS}) = (n+1-i)/n(n+1)/2, i = 1, \ldots, n \tag{3.105}$$

Rank reciprocal (RR) weights. This method uses the reciprocal of the ranks which are normalized by dividing each term by the sum of the reciprocals. The RR weights are defined by (Ibid.)

$$w_j(\mathrm{RR}) = 1/i \Big/ \sum_{j=1}^{n} \frac{1}{j}, \ \text{rank}\, i = 1, \ldots, n \tag{3.106}$$

For each of these methods, the corresponding weights for each rank, for numbers of criteria ranging from n = 2–10 are listed in Tables 3.32–3.37.

Table 3.32 (ROC) weights (Roberts and Goodwin, 2002)

Criteria				
Rank	*2*	*3*	*4*	*5*
1	0.7500	0.6111	0.5208	0.4567
2	0.2500	0.2778	0.2708	0.2567
3		0.1111	0.1458	0.1567
4			0.0625	0.0900
5				0.0400

Table 3.33 (ROC)

Criteria					
Rank	*6*	*7*	*8*	*9*	*10*
1	0.4083	0.3704	0.3397	0.3143	0.2929
2	0.2417	0.2276	0.2147	0.2032	0.1929
3	0.1583	0.1561	0.1522	0.1477	0.1429
4	0.1028	0.1085	0.1106	0.1106	0.1096
5	0.0611	0.0728	0.0793	0.0828	0.0846
6	0.0278	0.0442	0.0543	0.0606	0.0646
7		0.0204	0.0334	0.0421	0.0479
8			0.0156	0.0262	0.0336
9				0.0123	0.0211
10					0.0100

Table 3.34 (RS) weights (Roberts and Goodwin, 2002)

Criteria				
Rank	*2*	*3*	*4*	*5*
1	0.6667	0.5000	0.4000	0.3333
2	0.3333	0.3333	0.3000	0.2667
3		0.1667	0.2000	0.2000
4			0.1000	0.1333
5				0.0667

Table 3.35 (RS)

Criteria					
Rank	6	7	8	9	10
1	0.2857	0.2500	0.2222	0.2000	0.1818
2	0.2381	0.2143	0.1944	0.1778	0.1636
3	0.1905	0.1786	0.1667	0.1556	0.1455
4	0.1429	0.1429	0.1389	0.1333	0.1273
5	0.0952	0.1071	0.1111	0.1111	0.1091
6	0.0476	0.0714	0.0833	0.0889	0.0909
7		0.0357	0.0556	0.0667	0.0727
8			0.0278	0.0444	0.0545
9				0.0222	0.0364
10					0.0182

Table 3.36 (RR) weights (Roberts and Goodwin, 2002)

Criteria				
Rank	*2*	*3*	*4*	*5*
1	0.6667	0.5455	0.4800	0.4379
2	0.3333	0.2727	0.2400	0.2190
3		0.1818	0.1600	0.1460
4			0.1200	0.1095
5				0.0876

Table 3.37 (RR)

Criteria					
Rank	*6*	*7*	*8*	*9*	*10*
1	0.4082	0.3857	0.3679	0.3535	0.3414
2	0.2041	0.1928	0.1840	0.1767	0.1707
3	0.1361	0.1286	0.1226	0.1178	0.1138
4	0.1020	0.0964	0.0920	0.0884	0.0854
5	0.0816	0.0771	0.0736	0.0707	0.0682
6	0.0680	0.0643	0.0613	0.0589	0.0569
7		0.0551	0.0525	0.0505	0.0488
8			0.0460	0.0442	0.0427
9				0.0393	0.0379
10					0.0341

ROD is a weight approximation method that assumes that valid weights can be elicited through direct rating. In the direct rating method, the most important criterion is assigned a weight of 100, and the importance of the other criteria is then assessed relative to this benchmark. The 'raw' weights, (w_i^*) obtained are then normalized to sum to 1. Assuming that all criteria have some importance, this means that the ranges of the possible 'raw' weights will be

$$w_1^* = 100, \quad 0 \prec w_2^* \leq 100, \quad 0 \prec w_3^* \leq w_2^*$$

And in general:

$$0 \prec w_i^* \leq w_{i-1}^* (\text{where } i \neq 1) \tag{3.107}$$

These ranges can be approximated by representing all of the inequalities by less-than-or-equal-to expressions. The uncertainty about the 'true' weights can then be represented by assuming a uniform distribution for them. To determine ROD weights for general problems, it is needed to consider the probability distributions for the normalized weights that follow from the assumptions about the distributions of the raw weights. For $n > 2$, the density functions are a series of piecewise equations.

The means of each ROD for $n = 2$ to 10 have been found mathematically and are displayed in Table 3.38. For further information about the calculations behind, see Roberts and Goodwin (2002) (Table 3.39).

A graphical comparison of the ROD, ROC, and RS weights for nine criteria can be seen in Figure 3.19 (Roberts and Goodwin, 2002).

Table 3.38 ROD weights (Roberts and Goodwin, 2002)

Criteria				
Rank	*2*	*3*	*4*	*5*
1	0.6932	0.5232	0.4180	0.3471
2	0.3068	0.3240	0.2986	0.2686
3		0.1528	0.1912	0.1955
4			0.0922	0.1269
5				0.0619

Table 3.39 ROD

Criteria					
Rank	*6*	*7*	*8*	*9*	*10*
1	0.2966	0.2590	0.2292	0.2058	0.1867
2	0.2410	0.2174	0.1977	0.1808	0.1667
3	0.1884	0.1781	0.1672	0.1565	0.1466
4	0.1387	0.1406	0.1375	0.1332	0.1271
5	0.0908	0.1038	0.1084	0.1095	0.1081
6	0.0445	0.0679	0.0805	0.0867	0.0893
7		0.0334	0.0531	0.0644	0.0709
8			0.0263	0.0425	0.0527
9				0.0211	0.0349
10					0.0173

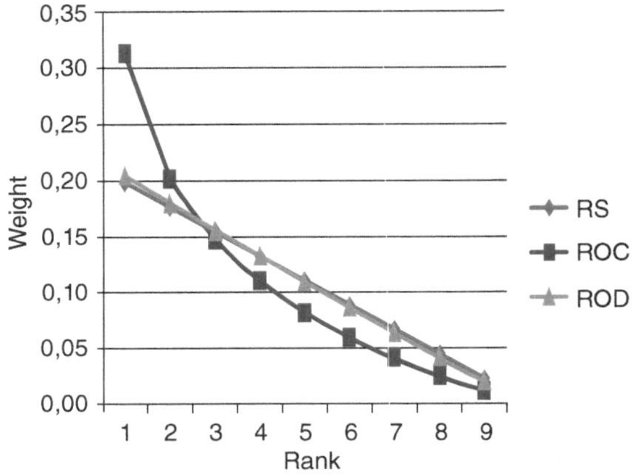

Figure 3.19 Comparison of weights for 9 attributes (Roberts and Goodwin, 2002).

There is a very close match between the ROD and RS weights. This matching is found whatever the number of criteria. Indeed, in general, the ROD weights tend toward the RS weights as the number of criteria increases. Thus, given that ROD weights are difficult to calculate when the number of attributes is large, a practical solution is to use RS weights for large criteria problems. The ROC weights depart markedly from both the RS and ROD weights.

The figure also demonstrates another benefit of using ROD instead of ROC weights. ROC weights are 'extreme' in that the ration of the highest to the lowest weights is so large that the lowest ranked criterion will only have a very marginal influence on the decision. In practice, criteria with relative importance as low as this would usually be eliminated from the decision model. The use of ROD weights goes some way to reducing this extreme value problem. However, it can be argued that the inclusion of criteria with very low weights, e.g. 0.02, does not contribute in any way to the overall result and therefore should be omitted from the analysis. For a discussion of this, see Barfod et al. (2011).

3.13.11 Pros and Cons of SMART

Pros. The structure of the SMART method is similar to that of the traditional CBA[30]in that the total "value" is calculated as a weighted sum of the impact scores. In the CBA, the unit prices act as weights and the "impacts scores" are the quantified (not normalized) CBA impacts. This close relationship to the well-accepted CBA method is appealing and makes the method easier to grasp for the decision maker.

Cons. In a screening phase where some poorly performing alternatives are rejected leaving a subset of alternatives to be considered in more detail, the SMART method is not always the right choice. This is because, as noted by

Hobbs and Meier (2000), SMART tends to oversimplify the problem if used as a screening method as the top few alternatives are often very similar. Rather, different weight profiles should be used, and alternatives that perform well under each different weight profile should be picked out for further analysis. This also helps to identify the most "robust" alternatives. The SMART method has rather high demands on the level of detail in input data. Value functions need to be assessed for each of the lowest level attributes, and weights should be given as trade-off.

In the SMART analysis, the direct rating method of selecting raw weights is normally used as it is cognitively simpler and therefore is assumed to yield more consistent and accurate judgments from the decision maker. These raw weights are then normalized, and this normalization process yields different theoretical distributions for the ranks. The means of these distributions are the ROD weights.

The formulae for the distribution of the ROD weights become progressively more complex as the number of criteria increases. Since the RS weights are so easy to calculate and closely match the ROD weights for higher numbers of criteria, it is recommended to use RS weights when working with problems involving large numbers of criteria, and in cases where it can be assumed that the appropriate alternative method for eliciting the 'true' weights would have been the direct rating method.

ORESTE.[31] Roubens originally developed this method in 1982. In this way, the ranking of alternatives is obtained using the rankings of the criteria and through their important functions.

The ORESTE Method algorithm is composed of three phases: definition of the input data (variants, criteria), construction of the complete global preorder of variants, and construction of the global partial preorder of variants with the application of indifference and conflict analysis (Mercier and Teghem, 1993). The final ranking of variants has a graphical character and corresponds to the final matrix constructed with the application of indifference (I), preference (P), and incomparability (R) relations (Roubens, 1982).

The ORESTE method was first introduced by Roubens (1982). ORESTE allows ranking the experiments in a complete order or in a partial order by considering incomparability. Many multi-criteria analysis methods require some detailed information concerning the different criteria, weights, order relation, preference functions, etc. However, in many real cases, it is quite difficult to provide this information (Givescu, 2007). ORESTE deals with the situation where the alternatives $A = \{a_1, a_2, \ldots, a_n\}$ are ranked according to criterion c_i ($i = 1$ to m), and the criteria themselves are ranked according to their importance. The objective of the method is to find a global preference structure on a set of alternatives A, which reflects the evaluation of alternatives on each criterion and the preference among the criteria. Since ORESTE only takes into account the ranking of alternatives and criteria, it is suited to problems with ordinal data. The method does not require quantification of criteria weights and alternatives' performance values, but only their ordinal rankings (Delhaye et al., 1991).

The ORESTE method has three assumptions:

1 Each criterion, K_i, induces a weak linear ordering on A.
2 K_i can be ordered according to their importance.
3 Incomparability and indifference measures, as defined below, can be selected.

Each alternative a_j is assigned a qualitative value, v_{ij}, relative to each K_i representing the rank value of its performance relative to K_i. A qualitative value, ω_i, is assigned to each K_i representing the rank value of its importance relative to the other K_i. On the basis of these values, the distance, $\delta(0, v_{ij})$, between an arbitrary origin, 0, and v_{ij} is defined such that for any two alternatives, a_e and a_f, and for each K_i

$$\delta(0, v_{ij}) \prec \delta(0, v_{if}) \text{ if and only if } v_{ie} \succ v_{if} \tag{3.108}$$

Distances are thus assigned to each alternative relative to each criterion on the basis of which a rank, $\rho(v_{ij})$, is assigned to each of the $m \times n v_{ij}$ with (Pastijin and Leysen, 1989)

$$1 \prec \rho(v_{ij}) \leq \text{mn} \tag{3.109}$$

where for any two alternatives, a_e and a_f, and for any two criteria, K_e and K_f,

$$\rho(v_{ee}) \leq \rho(v_{ff}) \text{ if and only if } \delta(0, v_{ff}) \tag{3.110}$$

A concordance index, C (a_e, a_f), is defined for each pair of alternatives a_e and a_f such that,

$$\begin{gathered} \sum v_{if} - v_{ie} \\ C(a_e, a_f) = i : \delta(0, v_{ie}) \succ \delta(0, v_{if}) \end{gathered} \tag{3.111}$$

The rank, $\rho(a_j)$, of an alternative a_j is defined by

$$\rho(a_j) = \sum_{i=1}^{n} v_{ij} \tag{3.112}$$

Quantitative values, γ and β, are defined where γ represents the maximal level and β represents the minimal level of incomparability between two alternatives (Bourguignon and Massort, 1994). A preference relation, $\prec$, is defined such that given any two alternatives, a_e and a_f, $a_e \prec a_f$ if and only if

$$\left[\frac{R(a_f) - R(a_e)}{n^2(m-1)} \leq \beta\right] \Lambda \frac{C(a_f, a_e)}{R(a_f) - R(a_e)} \leq \gamma_{\text{ef}} \tag{3.113}$$

PROMETHEE.[32] The PROMETHEE method is proposed by Brans, and it is a kind of method based on outranking relation between pairs of an alternative. The outranking method compares pairs of alternatives on each criterion first. The PROMETHEE method induces the preferential function to describe the preference difference between pairs of alternatives on each criterion. Thus, preference functions about the numerical difference between pairs of alternatives are built to describe the preference deference from the point of the decision maker's view. These functions' value ranges from 0 to 1. The bigger the function's value is, the difference of the preference becomes larger. When the value is zero, there is no preferential difference between the pair of an alternative. On the contrary, when the value is 1, one of the alternatives is strictly outranking the other.

Let $A_1, A_2, \ldots, A_m$ be m alternatives and $g_1, g_2, \ldots, g_n$ be n cardinal criteria and let y_{ij} be the criteria value of the i_{th} alternative A_i concerning the j_{th} criterion g_j. We will assume, without loss of generality, that all criteria are to be maximized.

We use $p_j\ (A_i, A_k)$ to denote the preference function on criterion g_j

$$P_j(A_i, A_k) = \begin{cases} 0 & y_{ij} \leq y_{kj} \\ P(y_{ij} - y_{kj}) & y_{ij} \succ y_{kj} \end{cases} \tag{3.114}$$

Brans and Mareschal's six such functions based on different criteria were introduced. They are the true criterion, quasi criterion, criterion with linear preference, level criterion, criterion with linear preference and indifference area, and gauss criterion.

Clearly, a different generalized criterion represents a different attitude toward preference structure and the intensity of preference. It was observed by Brans that the criterion with linear preference and indifference area had been mostly used by the user followed by gauss criterion for practical application. In both criteria, the intensity of preference changes gradually from 0 to 1, while in the other criteria, there are sudden changes in the intensity of preference.

Criterion with linear preference and indifference area is used. The preference function is defined as follows:

$$P_j(d_j) = \begin{cases} 0 & d_j \leq s_j \\ \left(|d_j| - s_j\right)/\left(r_j - s_j\right) & s_j \prec d_j \leq r_j \\ 1 & d_j \succ r_j \end{cases} \tag{3.115}$$

where $d_j = y_{ij} - y_{kj}$ denotes the preference difference between a pair of alternatives on criterion g_j.r_j and s_j are preference and indifference thresholds, respectively.

Using the preference function, we can get the preference degree for each pair of alternatives on each criterion. In order to get the overall preference degrees (A_i, A_k), the preference degree on each criterion should be aggregated by the following formula:

$$S(A_i, A_k) = \sum_{j=1}^{n} \omega_j P_j(A_i, A_k) \tag{3.116}$$

where ω_j represent the weight of the criterion g_j. In order to rank the alternatives from the best one to the worst one, the outgoing and incoming flows for each alternative are defined as follows:

The outgoing flow of alternative, A_i, is defined as follows:

$$\phi^+(A_i) = \sum_{A_k \in A} S(A_i, A_k) \tag{3.117}$$

The incoming flow of alternative, A_i, is defined as follows:

$$\phi^-(A_i) = \sum_{A_k \in A} S(A_k, A_i) \tag{3.118}$$

Based on the outgoing flow and incoming flow, the net flow $\phi(A_i)$ is defined by the following formula to represent the overall preference degree of alternatives $A_i\ and\ A_j$

$$\phi(A_i) = \phi^+(A_i) - \phi^-(A_i) \tag{3.119}$$

Alternatives can be ranked from the best to the worst one by the net flow. If $\phi(A_i) = \phi(A_j)$, the alternative A_i is indifferent to A_j; if $\phi(A_i) \succ \phi(A_j)$, the alternative A_i is indifferent to A_j.

3.13.12 Inferring the Weight of Criterion

This chapter is based on the hypothesis that the preference and indifference thresholds are known, but the weights of criteria are unknown. Additionally, the decision maker can give a preference relation of some alternatives. We assume that the preference relations of part of the alternatives set $A^T = \{a_1, a_2, \ldots, a_k\} \subset A, (k \prec m)$ are known. For each pair of the alternative a_i, a_k in A^T, one of the two following relations holds. $a_i S a_k, a_k S a_i \cdot a_i S a_k$ represents that decision maker prefers an alternative. In order to infer the weights of criteria, the following linear programming was constructed:

$$\begin{aligned} &\max \sum_{i=1}^{k} \sum_{j=i+1}^{k} e_{ij} \\ &s.t: \quad S(A_i, A_j) - e_{ij} \geq S(A_j, A_i) \end{aligned} \tag{3.120}$$

for all pairs of alternatives in A^T which satisfied the relation of $A_i S A_j$ $S(A_i, A_j) + e_{ij} \geq S(A_j, A_i)$ and for all pairs of alternatives in A^T which satisfied the relation of $A_j S A_i$.

$$e_{ij} \geq 0 \tag{3.121a}$$

In the aforementioned linear programming, there are $k(k-1)/2$ nonnegative decision variables. The goal function is a linear function of the unknown weights.

Thus, linear programming can be solved by the Lingo software. With the weight solved by the aforementioned linear programming, we can compute the net flow of the alternatives and rank the alternative from the best to the worst one.

The regime is a useful instrument to support a policy process. It offers a structure that helps to gather information on the different impacts of alternative policies. Using the software, the ranking process becomes easy. The most important difficulty is the determination of the weightsbecause it is very difficult to reach a consensus among the stakeholders on that and is mainly a political problem rather than a technical one.

REGIME. The REGIME method was developed in 1982 by Hinlopen and Nijkamp. In this method, the alternatives are compared in pairs and are assigned to superior alternative 1 and for the subtractive alternative of –1, and if none of the alternatives are superior to each other, a zero number is assigned to both of them. Then, the matrix obtained by comparing two alternatives is used for the superior alternative recommendation.

The REGIME method has been chosen to complement with the NAIADE method. This model is a discrete multi-criteria method that allows working with ordinal and cardinal data, and weighs as important coefficients of the criteria in the evaluation. For the REGIME description, Vreeker et al. (2002) are quoted here:

"Regime Analysis is a discrete multi-criteria method, and in particular, it is a Generalized"form of concordance analysis, based in essence on a generalization of pairwise comparison methods. In order to gain a better understanding of Regime Analysis, let us reiterate the basic principles of concordance analysis.

Concordance analysis is an evaluation method in which the basic idea is to rank a set of alternatives by means of their pairwise comparisons in relation to the chosen criteria. We consider a choice problem where we have a set of alternatives i and a set of criteria k. We begin our analysis by comparing alternative i with alternative k in relation to all criteria. After having done this, we select all criteria for which alternative $i$perform better than, or is equal to, alternative k. This class of criteria we call a "concordance set." Similarly, we define the class of criteria for which alternative i performs worse than, or is equal to, alternative k. This set of criteria is called a 'discordance set.'

Regime Analysis is able to examine both quantitative and cardinal data. In the case of choice problems with qualitative data, we first need to transform the qualitative data into cardinal data and then apply the Regime method. The Regime Software method is able to do so consistently. Due to this necessity, Regime Analysis is classified as an indirect method for qualitative data. This is an important positive feature. When we apply the marginalization of qualitative data through indirect methods such as the Regime Analysis, we do not lose information like indirect methods. This is due to the fact that in the direct methods, only the ordinal content of the available quantitative information is used.

To compare the alternatives, the sum of the weights of the criteria forming the discordance set is subtracted from the sum of the weights of the criteria forming the concordance set. This subtraction is called the net concordance index. The sign of this index is used to determine if one alternative is better than another in the pairwise comparison.

To rank, the alternative REGIME uses a probability distribution of the set of feasible weights as a performance indicator of the alternatives, in order to obtain a complete ranking of alternatives.

First of all, the impact matrix is constructed which indicates the performance of each alternative according to each of the chosen criteria. As well as ELECTRE and PROMETHEE, REGIME is a concordance method. Pairwise comparison between the set of alternatives according to each criterion is carried out. For each pair of alternatives *i* and *k*, the criteria are selected, for which alternative *i* is better or equal to alternative *j*. We call the set of these criteria concordance set. The criteria according to which an alternative *j* is worse or equal to alternative *i* are called discordance set. Then, the alternatives *i* and *j* are ranked by means of the concordance index C_{ij}, that is, the sum of the weights attached to the criteria according to which alternative *i* is better or equal to alternative *j*. Then, the concordance index C_{ji} is calculated, which is obtained summing up the weights of the criteria according to which alternative *j* is better or equal to alternative *i*. Finally, the net concordance index is calculated subtracting C_{ji} from C_{ji} ($\grave{\imath}ij = C_{ji} - C_{ji}$), which is positive if alternative *i* is preferred to alternative *j*. It must be noted that, since in most cases only ordinal information is available on the weights, but not trade-offs (we know that a criterion is most important than another one, but it is not known how much of a good performance of a criterion is sufficient to compensate a bad performance of another one), the net concordance index only tells whether an alternative is preferred to another, but not how much. Since sometimes it is not possible to obtain a complete ranking of the alternatives using only $\grave{\imath}ij$'s sign, a performance indicator p_{ij} is formulated for the criterion *i* with respect to the criterion *j*, which indicates the probability that an alternative is preferred to another one, that is, the net concordance index is positive. p_{ij} =prob ($\grave{\imath}ij$>0). Using the performance indicator, an aggregate probability index can be defined, which indicates the performance score. Next, we define an aggregate probability measure, which represents the performance score.

$$p_i = \frac{1}{I-1}\sum_{j \neq i} p_{ij} \tag{3.121b}$$

where *I* is the number of chosen alternatives. p_{ij} and p_i are estimated using a specific probability distribution of the set of feasible weights. More technical information can be described by Hinlopen et al.

The regime is a relatively easy method to use, provided that one can have access to a user-friendly software, like the one developed by the Free University of Amsterdam. It presents the same difficulties of many other MCA methods. The determination of alternatives, criteria, and weights entails a high degree of subjectivity although at the same time this subjectivity can also be made more explicit by the same implementation of the method.

The cost in terms of money and of time of an MCA analysis performed with Regime Analysis very much depends on the kind of criteria that are chosen – which increase the need for information– and on the data availability.

Once criteria are formulated and if the software is used and the information is available to operationalize those criteria, the ranking procedure is rather quick.

PAMSSEM.[33] In 1996, Martel invented this DM method. The basics of the PAMSSEM method are similar to those of PROMETHEE and ELECTRE, except that this method is also able to evaluate fuzzy criteria.

PAMSSEM is a hybrid outranking method that combines the key features of ELECTRE, PROMETHEE, and NAIADE methods. The outranking method is designed to deal with heterogeneous and missing data. As in ELECTRE III, PAMSSEM is making use of pseudo-criteria such as discrimination thresholds and considers two types of inter-criterion information: (1) a set of coefficients or weights that represent the relative importance between each criterion, and (2) veto thresholds between criteria when needed.

PAMSSEM computes a concordance, local discordance and outranking indexes for each pair of alternatives in the aggregation phase the same way as in ELECTRE. Once the concordance matrix is constructed, an outranking degree is computed for each pair of alternatives $a, b \in A$ represented by $\sigma(a,b)$. Similar to PROMETHEE, this measure defines to what degree alternative a is preferred over alternative b overall criteria globally.

The second and the last step in PAMSEEM is the exploitation phase, which is carried out exactly the same way as in PROMETHEE. A total preorder forced by computing a net outranking flow for each alternative as $\sigma(a) = \sigma^{+}(a) - \sigma^{-}(a)$. Finally, the alternatives are sorted in a decreasing order according to this measure in order to produce a final ranking.

Notes

1 Massachusetts Institute of Technology.
2 Multiple Attribute Decision Making.
3 Game theory is the formal study of conflict and cooperation. Game-theoretic concepts apply whenever the actions of several agents are interdependent. These agents may be individuals, groups, firms, or any combination of these. The concepts of game theory provide a language to formulate, structure, analyze, and understand strategic scenarios.
4 Multiple Criteria Decision Making. Problems may not always have a conclusive or unique solution (Hwang and Yoon, 1981).
5 Multiple Objective Decision Making.
6 Vector maximization (or minimization) problem.
7 Decision Maker.
8 Fuzzy logic is a complex mathematical method that allows solving difficult simulated problems with many inputs and output variables. Fuzzy logic is able to give results in the form of recommendation for a specific interval of output state, so it is essential that this mathematical method is strictly distinguished from the more familiar logics, such as Boolean algebra. This paper contains a basic overview of the principles of fuzzy logic.
9 Boolean algebra is a type of mathematical operation that, unlike regular algebra, works with binary digits (bits) 0 and 1. While 1 represents true, 0 represents false. Computers can perform simple to extremely complex operations with the use of Boolean algebra. Boolean algebra and Boolean operations are the basis for computer logic.

10 Simple Additive Weighting Method.
11 PROMETHEE (Preference Ranking Organization Method for Enrichment Evaluations) method is proposed by Brans and it is a kind of method based on outranking relation between pairs of alternatives. The outranking method compares pairs of alternatives on each criterion first. The PROMETHEE method induces the preferential function to describe the preference difference between pairs of alternatives on each criterion. Thus, preference functions about the numerical difference between pairs of alternatives are built to describe the preference deference from the point of the decision maker's view. These functions' value ranges from 0 to 1. The bigger the function's value is, the difference of the preference becomes larger. When the value is 0, there is no preferential difference between pair of alternatives. On the contrary, when the value is 1, one of the alternatives is strictly outranking the other.
12 Technique for Order Preference by Similarity.
13 Knowledge Management.
14 Elimination Et Choice Translation Reality.
15 Analytic Hierarchical Process.
16 For a matrix $\boldsymbol{A}$, a_{ij} denotes the entry in the i_{th} row and the j_{th} column of $\boldsymbol{A}$. For a vector $\boldsymbol{v}$, v_i denotes the i_{th} element of $\boldsymbol{v}$.
17 Decision Making Trial and Evaluation Laboratory.
18 **Maximum mean de-entropy algorithm.** As we mentioned earlier, the threshold value is determined by asking experts or by the researcher (as a decision maker). Choosing a consistent threshold value is time-consuming if the impact-relations maps are similar when threshold values are changed slightly. If we consider the total relation matrix as a partially ordered set, the order relation is decided by the influence value. The question about deciding a threshold value is equal to a real point set divided into two subsets. One subset provides information on the obvious interdependent relationships of factors but the relationships are considered not so obvious in another subset. The proposed algorithm is a way to choose the "cut point."

We propose the maximum mean de-entropy (MMDE) algorithm to find a threshold value for delineating the impact-relations map. In this algorithm, we use the approach of entropy, which has been widely applied in information science, but define another two information measures, de-entropy and mean de-entropy. In addition, the proposed algorithm mainly serves to search for the threshold value by nodes (or vertices). This algorithm differs from the traditional methods through which the threshold value is decided by searching a suitable impact-relations map. In this section, we use the symbol ν as the end of a definition or a step in the proposed algorithm.
19 Novel Approach to Imprecise Assessment and Decision Environments.
20 Evaluation of Mixed Data method.
21 Multiple attribute value theory.
22 Practically, all-important decisions involve analysis of several (or even many), typically conflicting, criteria. Analysis of trade-offs between criteria is difficult because such trade-offs for most problems are practically impossible to be defined a-priori even by analysts experienced in **Multi-Criteria Analysis** (MCA). Therefore, the trade-offs emerge during an interactive MCA which actually supports a learning process about the trade-offs. Hence, effective MCA methods are important for actual support of decision making processes, especially those related to policy-making.

IIASA has been developing novel methods for MCA since mid-1970s, and successfully applying them to many practical problems in various areas of applications. However, there are new practical problems for which the existing MCA methods (developed not only at IIASA but also by many researchers all over the world) are not satisfactory. In particular, discrete decision problems with a large number of criteria and alternatives (the latter making pairwise comparisons by the users impracticable) demand new methods. For example, MCA analysis of future energy technologies involves over 60 criteria and over 20 discrete alternatives; a careful requirement analysis of this application has proven that none of the existing MCA methods is suitable for an effective analysis of the corresponding problem. Moreover, this analysis has been done by a large number of stakeholders with diverse backgrounds and preferences; most of them have no analytical skills; therefore, the specification of preferences needed to be simple but still provide effective and intuitive analysis of the Pareto set.

The paper provides an overview of several new methods for MCA of discrete alternatives that have been implemented in the MCA, the web-based application for multiple criteria analysis of discrete alternatives.

23 Office of the Deputy Prime Minister.

24 UTilités Additives.

25 Multiple Attribute Utility Theory.

26 Web-HIPRE is a web-version of the HIPRE 3+ software for decision analytic problem structuring, multi-criteria evaluation, and prioritization.

27 Simple Multiple Attribute Rating Technique.

28 Multi-Criteria Decision Analysis, or MCDA, is a valuable tool that we can apply to many complex decisions. It is most applicable to solving problems that are characterized as a choice among alternatives. It has all the characteristics of a useful decision support tool. It helps us to focus on what is important, is logical and consistent, and is easy to use. At its core, MCDA is useful for:

- dividing the decision into smaller, more understandable parts;
- analyzing each part;
- integrating the parts to produce a meaningful solution.

When used for group decision making, MCDA helps groups talk about their decision opportunity (the problem to be solved) in a way that allows them to consider the values that each view is important. It also provides a unique ability for people to consider and talk about complex trade-offs among alternatives. In effect, it helps people think, rethink, query, adjust, decide, rethink some more, test, adjust, and finally decide.

MCDA problems comprise five components:

- goal;
- decision maker or group of decision makers with opinions (preferences);
- decision alternatives;
- evaluation criteria (interests);
- outcomes or consequences associated with alternative/interest combination.

29 Simple Multiple Attribute Rating Technique.

30 **Cost–Benefit Analysis (CBA)** is founded on the principles of welfare economics. It sorts out what the net benefits for the local population are by indicating which of the money flows in EIA are a cost and which are a benefit.

The data requirements to perform a CBA are extensive, and only a few studies have been found so far which actually applied CBA for evaluating sport events. Schaffer et al. presented a multiple account valuation of the costs and benefits of the 2010 Winter Games to counteract the grossly exaggerated claims of "over $10 billion in provincial GDP and more than 200,000 jobs" (p. 6) generated through a standard EIA. A CBA looks at the broader question of what society gains and loses as a result of staging an event.

ACBA needs to incorporate all costs and all benefits in order to determine whether there are any net benefits. On the cost side, the opportunity cost, and not the actual financial cost, must be taken into account. On the benefit side, the increase in value of consumption of local residents, including the public good value of the event and the consumer surplus, needs to be taken into account. One way to measure benefits is through willingness to pay valuation techniques.

31 Organisation, Rangement Et Synthese de donneés Relationnelles.

32 Preference Ranking Organization Method for Enrichment Evaluations.

33 Procédure d'Agrégation Multicritère de type Surclassement de Synthèse pour Évaluations Mixtes.

4 Risks in Supply Chain Management

4.1 Introduction to Supply Chain Management

Supply chain risk management (SCRM) focuses on developing new approaches for the management of disruptions. The field of SCRM has originated from the idea of Enterprise Risk Management (ERM), the paradigm for managing the portfolio of risks that threaten organizations (Gordon et al., 2009). It is a challenge to capture the multi-dimensional and inter-dependent behavior of the risks. Raw material passes through various processes, geographic and political regions, changes ownerships and modes of transportation before reaching the end customers in the form of the finished product (Handfield and Ernest, 2002; Stecke and Kumar, 2009). All of these processes expose potential points where supply chains are vulnerable to disruptions. Modern supply chain trends such as globalization, decentralization, outsourcing and Just-In-Time are introduced to try and make supply chains efficient. However, this has led to an increase in the number of exposure points (Stecke and Kumar, 2009). In order to identify these failure points within the network, supply chain systems need a holistic perspective to understand and capture the complex network of interconnected nodes. Complexity within the supply chain system can be defined as a condition occurring due to the association of numerous inter-related and inter-dependent entities in the supply system using several process inter-connections. System-oriented and holistic approaches to risk management are identified in the SCRM literature as important in complex, uncertain and volatile global environments (Harland et al., 2003; Oehmen et al., 2009; Tang and Nurmaya Musa, 2010). Systems thinking may provide a methodological and structured approach to risk management due to its ability to consider the systemic environments within the larger system. It is necessary to look at supply chain systems from a 'system of systems' perspective. Systems thinking supports in capturing the dynamic, complex and inter-dependent nature of the system (Sterman, 2000). This research intends to study the portfolio of supply chain risks through three distinctive phases such as concept development, implementation and evaluation.

Empirically grounded research is needed for setting practicable managerial guidelines for supply chain risk-related problems (Juttner et al., 2003). From the literature survey on SCRM, the qualitative research approach has been widely used with several empirical studies and conceptual models in early research.

Various algorithm-based quantitative modeling techniques (Nagurney et al., 2005; Towill, 2005; Yang et al., 2005; Goh et al., 2007; Wagner and Neshat, 2010) have been effectively used in the past to solve supply chain network disruption problems.

Modern nature-inspired evolutionary algorithms have been used more recently for solving large, dynamic and complex optimization problems (Chiong, 2009). Different interdisciplinary theories like real options (Hult and Craighead, 2010), game theory (Xiao and Yang, 2008) and simulation (Kim et al., 2006; Wu and Olson, 2008) have shown some potential for managing supply chain disruptions. Several qualitative as well as quantitative research methods are utilized in the SCRM context. However, these important studies have either looked at risks across a dyad or one risk at a time. These studies do not provide a methodology of considering the influence of multiple risks on a supply system network, nor do they suggest a methodology for depicting risk propagation. The research reported in this chapter provides practitioners as well as researchers an approach to consider multiple supply chain risks and to capture their behavior over a period in supply chain network. The holistic risk management framework, systematic research design process and quantitative supply chain risk modeling bring together a unique capability for capturing the overall behavioral performance of risks. More recently, a system-oriented approach has been identified to be promising for modeling complex and dynamic problems (Cheng and Kam, 2008; Oehmen et al., 2009; Tang and Nurmaya Musa, 2010; Sheffield et al., 2012). The research intends to add to the existing work in SCRM by developing a holistic, systematic and quantitative risk assessment approach for measuring the overall risk behavior.

The paper is structured as follows: a brief literature review on SCRM within the context of ERM is discussed in the next section. Principles of systems thinking are utilized to build the framework for SCRM in Section 3. Section 4 describes the research design implemented for the risk assessment process along with the data collection activity for the research. Risk attributes are modeled based on the supply chain risk model in Section 5. System-based risk modeling is attempted through statistical and simulation modeling and is based on the developed supply chain risk model. Section 6 draws important insights from the conducted risk assessment to investigate overall risk behavior. Finally, the paper concludes with a discussion on important research contributions for practitioners as well as researchers from operational as well as strategic perspectives.

4.2 Literature Review

SCRM is a crucial and fundamental element of ERM addressing the supply side, even though SCRM and ERM are often perceived as separate functions within the firm (Blome and Schoenherr, 2011). SCRM, to a certain extent, can be compared to project and/or enterprise risk management as both environments consist of several nodes of network interconnected and working together for a single objective. According to Kleindorfer and Saad (2005) historically, operations or disruption risk management has been included under Integrated ERM. Hence, the approach to modeling of risks is built on the principles of ERM and SCRM literature. "*Enterprise risk management is defined as a process applied in terms of strategy setting across the enterprise, designed to identify and manage potential events that may affect the organization to provide reasonable assurance regarding the achievement of set objectives*" (COSO, 2004). The aligning link between ERM and SCRM processes has received very limited attention in the existing research (Blome and Schoenherr, 2011) but the research on risk management has evolved into numerous distinctive fields like financial risk management, healthcare risk management, project risk management, SCRM, etc. (Harland et al., 2003; Handfield and McCormack,2007).

Although supply chain risks are discussed significantly within the SCRM literature, there is limited information on how to deal with them from a practical perspective on short-term as well as long-term basis (Blackhurst et al., 2005). The attention given to assessing supply chain risks is fairly limited (Rao and Goldsby, 2009). Researchers suggest that an approach to risk management needs to follow a formal and structured process (Khan et al., 2008). Colicchia and Strozzi (2012) predict a need for a comprehensive and dynamic approach to SCRM. Identifying risks is the first step in developing efficient risk management procedure. It is evident from a systematic literature review on SCRM that qualitative as well as quantitative research methods are utilized for solving supply chain issues. Conceptual as well as empirical methods along with the case study-based approach are found to be commonly used. Quantitative tools like mathematical modeling and simulation techniques have recently been used to understand the intricacies of the SCRM field. Systems thinking and system dynamics (SD) can be effectively used for holistically studying different risk issues within a supply chain network. Although there are some instances of studying supply chain risks using the systems approach, in general, we find that systems thinking-based research approaches are largely unexplored for solving SCRM problems. Supply chains could be benefited by developing models that are able to model the risks from complex and dynamic networks (Stecke and Kumar, 2009). Identifying the points of failure by developing dynamic models to capture vulnerability in the supply chain would benefit researchers and practitioners for proactively mitigating the risks. Hence, a thorough investigation of supply chain risks for understanding their complex phenomenon is essential.

Risk management is increasingly becoming an integral part of a holistic SCM design (Christopher and Lee, 2004). SCRM follows a fairly traditional risk management process but is driven by the systemic interrelationships focused at identifying and reducing risks not only at an organization level but the entire supply chain. In general, SCRM consists of management processes such as identification, assessment, mitigation and control of risks (Hallikas et al., 2004). Risk classification and identification has been exhaustively discussed in the SCRM literature. Wold and Shriver (1997) define risk assessment as the process of analyzing the vulnerability to threats and recommending solutions to reduce the level of risk to an organization. The risk assessment process thus covers the most critical function of risk management. Chaudhuri et al. (2013) suggest that the assessment of supply chain risks should start during the new product development process due to the growing uncertainty in supply chains. Multidisciplinary approaches have been attempted for building models for supply chain risk analysis in the literature. Wu et al. (2006) and Wang et al. (2012) use analytical hierarchy process to model supply chain risk assessment. Multi-stage influence diagram (Liu, 2009), Monte Carlo approach (Klibi and Martel, 2012), interpretive structural modeling (Diabat et al., 2012; Hinlopen et al., 1983), partial least square method (Kern et al., 2012) and several other methods from MS/OR (Bryson et al., 2002) have been utilized by academics to test models for supply chain risk assessment. Nevertheless, risk assessment in supply chains is bounded by operational and economic constraints for a detailed study (Pai et al., 2003; Liu et al., 2009; Martel and Matarazzo, 2005). According to a leading multinational consultancy service firm, the risk assessment in industry setting is conducted based on previous experience and forward-thinking analysis is a must for effective risk mitigation (PricewaterhouseCoopers, 2008). Moreover, supply chain risk behavioral performance is inherently unpredictable and chaotic. Hence, supply chain practitioners demand a vigorous risk assessment mechanism to protect organizations against unforeseen disruptive events. Proactive assessment and execution are a key consideration for robust SCRM (Sodhi and Tang, 2012). The research attempts to bridge the gap between theory and practice in using a systems perspective within SCRM by developing a robust, systemic risk assessment methodology (Tables 4.1 and 4.2).

Table 4.1 A summary of the working definitions

Concept	*Working definition*
Supply chain	A system of suppliers, manufacturers, distributors, retailers and customers in which material, financial and information flows connect participants in both directions (Fiala, 2005)
Supply chain management	The function responsible for the transport and storage of materials on their journey from original suppliers through intermediate operations to final customers (Waters, 2007)
Risk	A threat that something might happen to disrupt normal activities and stop things happening as planned (Waters, 2007)
Supply chain risk management; the management of risk in supply chains	An umbrella concept incorporating the identification, analysis and control of risk. It refers to the overall function responsible for all aspects of risk to the supply chain; it ensures that the principles established by the senior managers are applied to logistics risk (adapted from Waters, 2007)
Risk identification	The initial step of supply chain risk management in which potential threats to the chain are identified (Waters, 2007)
Risk analysis	The second step of supply chain risk management in which the risks are evaluated and assessed (Waters, 2007)
Risk control; risk management action; risk mitigation; risk response	The third step of supply chain risk management, referring to actual risk management actions based on the information produced during the identification and analysis stages (adapted from Waters, 2007)

(*Continued*)

Concept	Working definition
Multimodal supply chains	International transport systems combining various modes of transport, such as ship, rail and truck, primarily through the use of containers (Beresford et al., 2011)
Supply chain visibility	The actors' knowledge of what goes on in other parts of the chain (Christopher and Lee, 2004)
Supply chain collaboration	Process-focused or relationship-focused collaboration: the former is viewed as a business process whereby two or more supply chain partners work together toward common goals, whereas the latter refers to the formation of close, long-term partnerships among supply chain members working together and sharing information, resources and risk in order to achieve mutual objectives (Mentzer et al., 2001; Stank et al., 2001; Manthou et al., 2004; Sheu et al., 2006)
Supply chain information exchange	The extent to which information is communicated between the partners in the supply chain (Fiala, 2005; Lysons and Farrington, 2006)

Table 4.2 Different perceptions of supply chain risk (adapted from Sodhi et al., 2012)

Author	Scope of risks
Jüttner et al. (2003)	Environmental sources, network sources, and organizational sources
Spekman and Davis (2004)	(1) inbound supply, (2) information flow, (3) financial flow, (4) the security of a firm's internal information system, (5) relationship with partners, and (6) corporate social responsibility
Cavinato (2004)	(1) physical, (2) financial, (3) informational, (4) relational, and (5) innovational sources
Chopra and Sodhi (2004)	Categorize supply chain risks at a high level as disruptions or delays. These risks pertain to (1) systems, (2) forecasts, (3) intellectual property, (4) receivables, (5) inventories and (6) capacity
Christopher and Peck (2004)	(1) process, (2) control, (3) demand, (4) supply, and (5) the environmental
Kleindorfer and Saad (2005)	Risks sources and vulnerabilities from (1) operational contingencies, (2) natural hazards, and (3) terrorism and political instability
Bogataj and Bogataj (2007)	(1) supply risks, (2) process risks, (3) demand risks, and (4) control risks
Sodhi and Lee (2007)	(1) supply, (2) demand, and (3) contextual risks requiring both strategic and operational decisions
Tang and Tomlin (2008)	(1) supply, (2) process, and (3) demand risks, (4) intellectual property risks, (5) behavioral risks and (6) political/social risks
Manuj and Mentzer (2008a)	(1) supply, (2) operations, (3) demand, and (4) other risks including security and those related to currency
Manuj and Mentzer (2008b)	(1) supply, (2) operational, (3) demand, (4) security, (5) macro, (6) policy, (7) competitive, and (8) resource risks
Oke and Gopalakrishnan (2009)	Consider low-impact-high-frequency and high-impact-low-frequency risks in three major categories: (1) supply, (2) demand, and (3) miscellaneous
Rao and Goldsby (2009)	(1) framework, (2) problem-specific and (3) decision-making risk

Xia and Chen (2011) identify four different forms of risk impact related to quantity, cost, quality and time. The risk analysis in the present study included the three latter forms as identified by the actors in the field. Furthermore, Manuj and Mentzer's (2008a) risk classification presented in Table 4.3 was used as a starting point in determining the source categories. Also, you can see Appendix B about main issues discussed over the years.

This research reflects the notion of conceptual causality introduced by Waters (2007). The following figure illustrates the focal position of risk management in the holistic management of the uncertainty-driven risks confronting the supply chain (Figure 4.1).

Table 4.3 Risk classification (Manuj and Mentzer, 2008b)

Type of risk	*Source*
Supply risks	Disruption of supply, inventory, schedules, and technology access; price escalation; quality issues; technology uncertainty; product complexity; frequency of material design changes
Operational risks	Breakdown of operations; inadequate manufacturing or processing capability; high levels of process variations; changes in technology; changes in operating exposure
Demand risks	New product introductions; variations in demand (fads, seasonality, and new product introductions by competitors); chaos in the system (the Bullwhip Effect on demand distortion and amplification)
Security risks	Information systems security; infrastructure security; freight breaches from terrorism, vandalism, crime, and sabotage
Macro risks	Economic shifts in wage rates, interest rates, exchange rates, and prices
Policy risks	Actions of national governments such as quota restrictions or sanctions
Competitive risks	Lack of history about competitor activities and moves
Resource risks	Unanticipated resource requirements

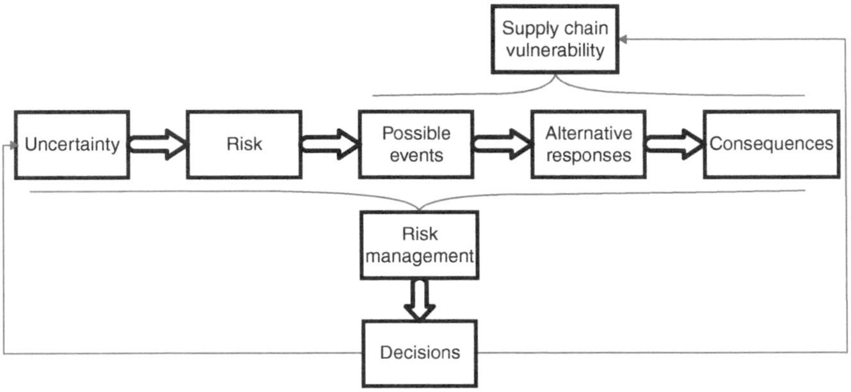

Figure 4.1 The connections between the various concepts covering risk management in supply chains (Waters, 2007).

4.3 Approach to Defining/Classifying Supply Chain Risk

The terms 'risk', 'uncertainty', 'disruption' and 'disaster' are frequently and interchangeably used in supply chains to describe the perceptions and interpretations of individuals and organizations. A general interpretation of risk is influenced by the negative consequences of variation in expected outcomes, their impact and likelihoods (March and Shapira, 1987). Risk events are also studied with core supply chain activities and investigated with common business practices. Christopher and Peck (2004) relate the risks with the vulnerability and likelihood of being lost or damaged. Interruptions to the flow of information, material and finance from the original supplier to the end user which cause a mismatch between demand and supply are also considered as risks (Juttner et al., 2003).

In line with the definitions discussed earlier and to relate the risks with supply chain functional aspects, we categorize the orientation of risk definitions related to operational characteristics, market characteristics, business/strategic characteristics, product characteristics and others. Table 4.4 shows the risk characteristics and features in each of the categories.

Table 4.4 Risk definition criterion and description

Risk definition criterion	*Definition description/ Characteristics*	*Risk issues*
Related to operational characteristics	Operational features of supply chain which mismatch demand and supply or even disrupt the functioning of supply chain and interrupt the flow of material, product and information	Supply disruptions, demand uncertainties, machine/system failures, improper planning and execution, information and security risks
Related to market characteristics	Market fluctuations which cannot be predicted precisely and change their nature, impact and occurrence over time.	Price variability, customer behavior and expectations, competitor moves, exchange rates, environmental risks and disasters
Related to business/ strategic characteristics	Specific characteristics of business, sector, their strategies and environment which cause an undesired event to happen and negatively affect the supply chain performance	adverse effects of strategies such as outsourcing, single sourcing, lean manufacturing, improper supply network design, forecasting errors, lack of coordination and information sharing
Related to product characteristics	Features related to the specific nature of products which make the supply chain vulnerable to risk and uncertainties	Short product life cycles, complexity in product design and manufacturing, desire for variety of products, need for multifunctional products
Miscellaneous	Various other characteristics can also be considered which may fit in the above-mentioned category or can be studied separately	political risks, credibility risks brand image risk, social risks, ecological risks, etc.

4.4 Risk Issues Related to Structural Elements of the Supply Chain

Supply chain structures are complex networks of different players (including lower tier suppliers to the end customer) established with core objectives to minimize the costs, maximize the value and explore new markets through effectively managed relationships among members (Hallikas et al., 2002; Blackhurst et al., 2007; Trkman and McCormack, 2009; Tuncel and Alpan, 2010). Though networking is a way to take advantage of collaboration and partnership among various supply chain players, it becomes a source as well as a medium through which risks are generated and propagated to the entire network. To capture the structural dimension of the supply chain risks, we classify the literature for the perspectives of upstream[1] and downstream[2] (Table 4.7). We also study the literature with a single focal firm point of view but observe that most of the risk issues related to a single firm are more relevant in a dyadic frame. Therefore, we prefer to analyze the risk issues from a relational point of view in the form of dyads. To provide deeper insights into the upstream risks, we further classify them considering the elements of supply system design: number of suppliers (single/multiple sourcing), location of suppliers (local/global sourcing) and coordination and information sharing, and thus divide the literature into supply system design and coordination and information sharing. Other issues such as supplier behavior, traits, etc., are considered under the general issue's category.

Downstream risks usually relate to the fluctuations in demand, volatile market conditions, customer behavior, technological changes and shorter product life cycles. At one end, these risks are associated with the physical distribution and product flow toward the downstream side and on the other hand, they are related to forecasting issues (Szwejczewski et al., 2008). These risks are usually the outcome of a mismatch between actual demand and projected demand resulting in a demand and supply mismatch throughout the supply chain. We focus on two discriminating elements and classify the demand issues as market volatility and demand fluctuation and coordination and information sharing.

4.5 Level of Implementation of Risk Management Approach

Implementation of SCRM is an extremely critical task requiring a sound knowledge of business functions, market trends and financial and infrastructural status of the organization as well as the entire supply chain. Implementation of SCRM generally requires three steps given as: identifying the potential risks to the organization, assessing the risks and aftermaths and adopting suitable risk managing strategies. A hierarchy exists between these phases and the higher phase subsumes the lower phase (Dailun, 2004).

Risk identification is an important first step in any risk management effort. Numerous approaches have been proposed to identify the risks in supply chains, classified as: the common listing approach, where the analysis of historical events is utilized to gain insight into future risks; taxonomy-based approaches, which provide a consistent framework to elicit and organize risk identification activities related to various business functions; scenario analysis, in which key risk factors and their effects on supply chain performance are analyzed to develop a risk profile, making it easy to develop contingency plans at the operational level; risk mapping, with the capability of exposing the vulnerability of supply chains to potential risk before their occurrence.

Assessing the risks qualitatively or quantitatively is an essential task after the risk identification. When sufficient past data and expertise are available, quantification of risks is meaningful; otherwise, qualitative methods are more appropriate. We categorize the methods as assessing the risk sources and risk characterization, with the latter being more rigorous. Assessing the sources and exposure is effective when limited past data is available. The sources of risks and exposure are evaluated and subjectively indexed/ranked based on the assessor's perspective and experience. Risk characterization provides a broader framework for risk assessment, grouping and prioritizing employing analytical models.

Various strategic and operational risk management stances are reported in the literature. We classify them as the shaper, accepter and recovery approach. In the shaper approach, attempts are made to shape (reduce the impact and frequency) the uncertainty factors without changing the existing settings of the supply chain, while in the accepter approach, risks are accepted and supply chains are reinvestigated and redesigned. Recovery strategies mainly support quick recovery mechanisms after severe damage in the supply chains.

4.6 Framework for Supply Chain Risk Management

In this section, the conceptual framework for SCRM is developed using a systems perspective. The conceptual framework follows standard risk management processes; risk identification, risk assessment and risk mitigation as seen in Figure 4.2. Although the processes may look similar to standard risk management, the difference lies in the approach to the problem and the research methodology implemented for the study.

The systematic development of the framework was achieved beginning with a standard risk management process. To capture the intricacies involved in each process, two stages were developed for each process during data experimentation. Each stage in the conceptual framework was improved through a continuous feedback loop system. Risk taxonomy is the first stage in the framework where the risks are identified and classified from the pool of risks. Risks trending, the second stage in the risk identification process is for predicting the operational boundaries of the risk variables. The risk assessment process is the major focus of our research and hence discussed exhaustively in this chapter. A research design for assessing the dynamic behavior of risks is developed. For risk modeling, a model is developed in order to capture the impact in terms of cost and time (delay) and the possible failure point due to disruption. Risk modeling and sensitivity analysis stages in the risk assessment process are attempted

Figure 4.2 Framework for supply chain risk management.

through quantitative modeling techniques to evaluate the overall performance of the risks. The risk mitigation process is classified into two stages as strategic planning and risk mitigation. Strong inferences drawn from risk trending, risk modeling and sensitivity analysis provide directions for the risk mitigation. New risk mitigation strategies identified from the study are utilized for future projects. The framework for SCRM forms a closed-loop system for continuous improvement. The systematically developed framework for SCRM (Figure 4.2) is believed to capture the overall nature of risks through a structured study discussed in the later part of this research. All the activities described in the conceptual framework are structurally followed for modeling supply chain risks.

Intensifying competition since the 1990s has forced companies to improve efficiency in many aspects of their business. While getting rid of the 'slack' in their supply chains, they expose themselves to greater uncertainty, and this is what SCRM aims to control. As a concept, it is at the intersection of supply chain management and risk management (see Figure 4.3). The term "supply chain management" is still relatively new, first appearing in the literature in 1982 (Keith and Webber, 1982). It was originally used in the context of logistics, and emphasized a reduction in inventories within and across organizations (Cooper et al., 1997). The concept, in general, is still new, and in many companies, it is unknown (Blos et al. 2009).

Lavastre et al. (2011, p. 8) define SCRM as "the management of risk that implies both strategic and operational horizons for long-term and short-term assessment". According to Brindley (2004), it means "the management of supply chain risk through coordination or collaboration among supply chain partners so as to ensure profitability and continuity". The aim therefore is to control the

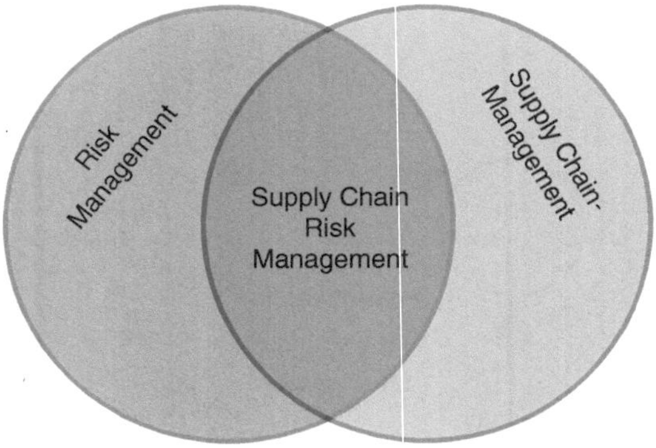

Figure 4.3 Supply chain risk management.

risks and uncertainties caused by, or impacted on, logistics-related activities or resources (Waters, 2007). It is executed collaboratively with partners in a supply chain by applying risk-management-process tools (Norrman and Lindroth, 2002). According to Christopher et al. (2011), companies managing risks in a global economy should adopt a multidisciplinary approach.

SCRM starts from the identification and computation of probable risks and their possible impact on operations in the supply process. The first stage is to identify the direct risks to its operations, and then to consider the potential causes of risk at every significant link in every step of the chain (Lysons and Farrington, 2006). A further aim is to identify the potential sources of risk and implement appropriate actions to avoid or contain vulnerability.

According to Tang (2006), there are two forms of governance in SCRM, namely coordination and collaboration. Along the continuum of supply chain relationships presented earlier in Figure 4.3, it is positioned between cooperation and hierarchy, depending on the depth of relations and information exchange and the form of governance with respect to the focal company.

According to Jüttner et al. (2003), SCRM comprises four main elements: (1) assessing the risk sources, (2) identifying the concepts, (3) tracking the drivers and (4) mitigating the risks. Kleindorfer and Saad (2005), in turn, propose three process elements, namely (1) specifying the sources of risk and vulnerability, (2) assessment and (3) mitigation, which is fairly close to what Waters (2007) proposes (see Figure 4.6). Sodhi et al. (2012) identify similar elements from the literature, differentiating between (1) risk identification, (2) risk assessment, (3) risk mitigation and (4) responsiveness to risk (either operational or catastrophic). Hallikas et al. (2004) propose a similar model: (1) risk identification, (2) risk assessment, (3) risk management action and (4) risk monitoring.

SCRM could thus be viewed as a strategic-management activity given that it can affect the operational, market and financial performance of firms (Narasimhan and Talluri, 2009). Figure 4.4 presents a framework for risk management in supply chains.

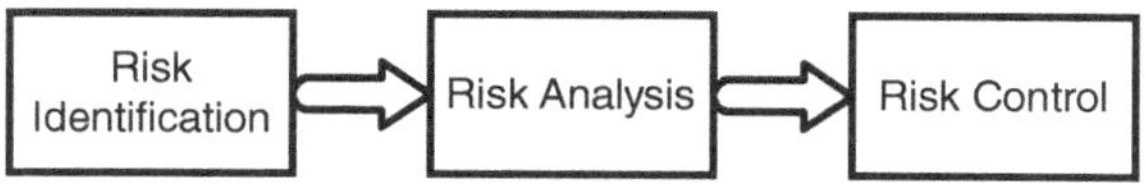

Figure 4.4 A framework for managing risks in supply chains (adapted from Waters, 2007).

4.7 Risk Identification

It is generally agreed that identification is the initial step in the process of SCRM. According to Waters (2007), identifying the risks is a key activity on which all other aspects of the process are based. However, in reality, it is virtually impossible to list every conceivable risk, and identification will only cover the most significant in terms of their effect on the supply chain. Inter-organizational actors usually have the most intimate knowledge of the organization and its conditions, but do not necessarily have the capability to identify risks. Organizations cannot rely on personal knowledge and informal procedures, but need some formal arrangements (Waters, 2007).

There are numerous techniques covering the management of risk in supply chains (see, e.g., Peck et al., 2003). According to Lavastre et al. (2011), the tools tend to be used more in the risk identification and assessment phases. One of the most popular Failure Mode and Effects Analysis (FMEA) is introduced in this study. A proactive tool for risk identification and analysis, FMEA was developed by NASA in 1963 to identify, evaluate and prevent product and/or process failures (Hu et al., 2009). It is considered a powerful and effective analytical tool for examining possible failure modes in a system (Chen, 2007). According to Van Leeuwen et al. (2009), FMEA can be used to prioritize risks and monitor the effectiveness of risk-control activities, and therefore is valuable in terms of identifying risks, including those related to human factors. The conventional FMEA procedure suffers from inadequate definition of some steps, high uncertainty and even decision-making failures throughout the procedure (Bluvband and Grabov, 2009). Further disadvantages include the assumption that the failure modes are all single events and level in nature; the fact that it requires a lot of time, resources and cooperation to achieve the required detail; that it takes limited account of human error; and that it may give an identical risk-priority value to different events even though the implication may be totally different (Pillay and Wang, 2003; Rhee and Ishii, 2003; Hsu et al., 2011; Xiao et al. 2011). Scholars acknowledge the limitations of FMEA, and there have been attempts to overcome some of the drawbacks (see, e.g., Franceschini and Galetto, 2001; Sankar and Prabhu, 2001; Arunachalam and Jegadheesan, 2006; Chen, 2007; Wang et al., 2009).

4.8 Risk Analysis

Analyses of risks affecting the supply chain should also take into account where the risks derive from so that contingencies can be built in to mitigate their effects

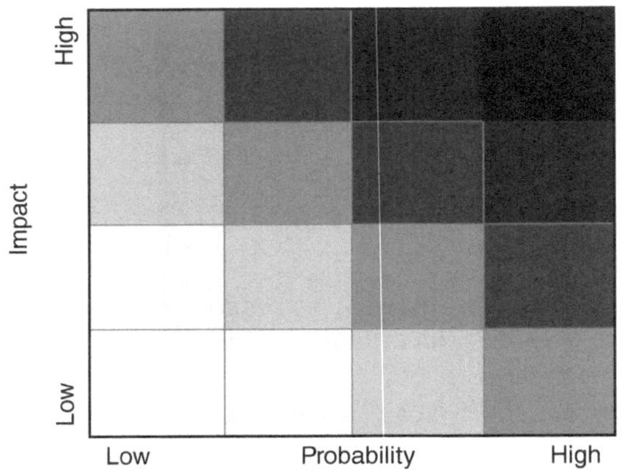

Figure 4.5 The risk matrix (adapted from Norrman and Lindroth, 2002).

or prevent their realization. Handfield and McCormack (2008) define the severity of disruption as the number of nodes within a supply network whose ability to ship and/or receive goods and materials has been affected by an unplanned, unanticipated event. All supply chains carry some risk, but the extent depends on multiple factors including the density, criticality and node density of the network. Norrman and Lindroth (2002) developed a two-dimensional risk matrix based on measures of impact and probability. The matrix can be used to assess the severity of risk, as illustrated in Figure 4.5.

Lindroth and Norrman (2004) further propose a three-dimensional framework (see Figure 4.6 below) comprising the risk-handling focus, the type of risk and the unit of analysis. The framework is useful for examining the multidimensional construct of risk, which many authors recognize (Zsidisin, 2003; Peck 2005).

4.9 Risk Control

Consequent to their analysis, the risks have to be properly managed (Gerber and von Solms, 2005). According to Scarff et al. (1993, p. 2), the management of risk refers to the "overall process by which risks are analyzed and managed", whereas risk management entails "planning, monitoring and controlling activities which are based on information produced by risk analysis activity". In order to avoid confusion among these two terms, the concept of SCRM is assumed in this study to include the overall process in which risks are identified, analyzed and controlled, whereas the concept of risk control refers to the actual risk management actions based on the information produced from risk identification and analysis.

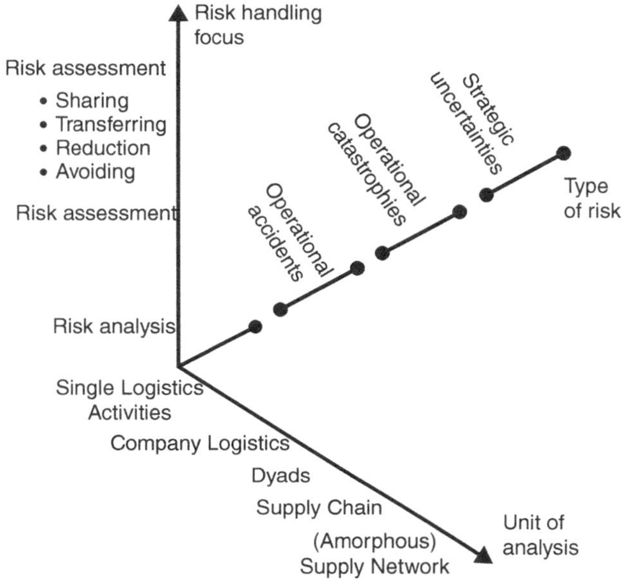

Figure 4.6 A framework for assessing and positioning risk in supply chains (Lindroth and Norrman, 2001).

Waters (2007) defines the activity of SCRM following the risk analysis as "designing an appropriate response", in other words determining the most appropriate way of dealing with the risks. Once they have been identified and prioritized, and the amount of attention each risk deserves has been assessed, careful consideration should be given to the amount of resources required to deal with them. Waters (2007) suggests the following range of responses to risk: ignore or accept it, reduce the probability, reduce or limit the consequences, transfer, share or deflect the risk, make contingency plans, adapt to it, oppose a change or move to another environment. Tummala and Schoenherr (2011) present a list of 'risk triggers', and like many other authors divide the risks according to their 'consequence severity level' and 'risk probability', which finally determines the severity.

4.10 Research Design

The research design implemented for the risk assessment is based on the application of systems thinking concepts. Sterman (2000, p. 4) defines systems thinking as *"the ability to see the world as a complex system, in which we understand that you can't just do one thing and that everything is connected to everything else"*. Luna-Reyes and Andersen (2003) define systems thinking as a modeling approach used for conceptualizing and analyzing interdependencies of the system. Sterman (2000, p. 4) has suggested that when one is a part of a complex system, it is difficult to learn about it. SD is thus a *"method to enhance learning in complex systems"* (Sterman, 2000, p. 4) and systems thinking is crucial during the system conceptualization phase in SD (Forrester, 1961). The systems thinking approach provides a structured development process from conceptualization to the end of system lifecycle (Forrester, 1961, 1994; Sterman, 2000). Qualitative as well as quantitative data can be used for conceptualizing and modeling the system (Luna-Reyes and Andersen, 2003). Tools like simulation/system dynamics and different algorithm modeling have the potential to capture static as well as the dynamic behavior of supply chains. Following the systems thinking approach, a step-by-step experimental research design for risk assessment is developed and implemented in this section.

The research design for modeling supply chain risks primarily focuses on the risk assessment process in the proposed framework for SCRM. Empirical research designs use statistical analysis, OR modeling and simulation techniques to draw the results (Luna-Reyes and Andersen, 2003).

Figure 4.7 shows the developed research design for modeling supply chain risks. It implements two distinctive approaches for evaluating the complete risk behavioral performance. The left side is termed as 'statistical approach' for behavioral risk assessment, and the right side is termed as 'systems approach' for exploring the risk performance. Both modeling platforms run parallel to each other during the risk assessment process and are later combined to extract comprehensive results.

A reputed Aerospace and Defense organization in the UK was approached for modeling of the supply chain risks phenomenon. The organization has its supply chain network spread across the world. The typical nature of supply chain activities for this organization involves design, manufacture, delivery and after sales maintenance of the product. Several informal meetings were held with the organization to discuss the identified research problem and to further understand the gap in modeling supply chain risks from an industry perspective.

The discussions identified that for the case company, there was a need to move from a traditional risk management to an enhanced risk management approach. The knowledge gap in relating SCRM theory to industry practice was another important concern raised by this collaborating organization. By matching the research objectives, the collaborative project generated the ideal platform for participatory experimental research while working with the Risk Manager and System Engineers from the organization. Qualitative as well as quantitative data was collated from different internal projects within the organization. The project inherently was a product development environment representing a global supply chain network. The data collection for an experimental research can be in a wide variety of formats. This can be in the form of documents, reports, registers, spread sheets, audio/video recordings, etc. The data for this research was in the form of risk documentation in the risk register. The quantitative risk register data was supported with qualitative data in the form of informal interviews and secondary data made available from company reports and Internet sources. Initially, the project risk data was thoroughly studied and

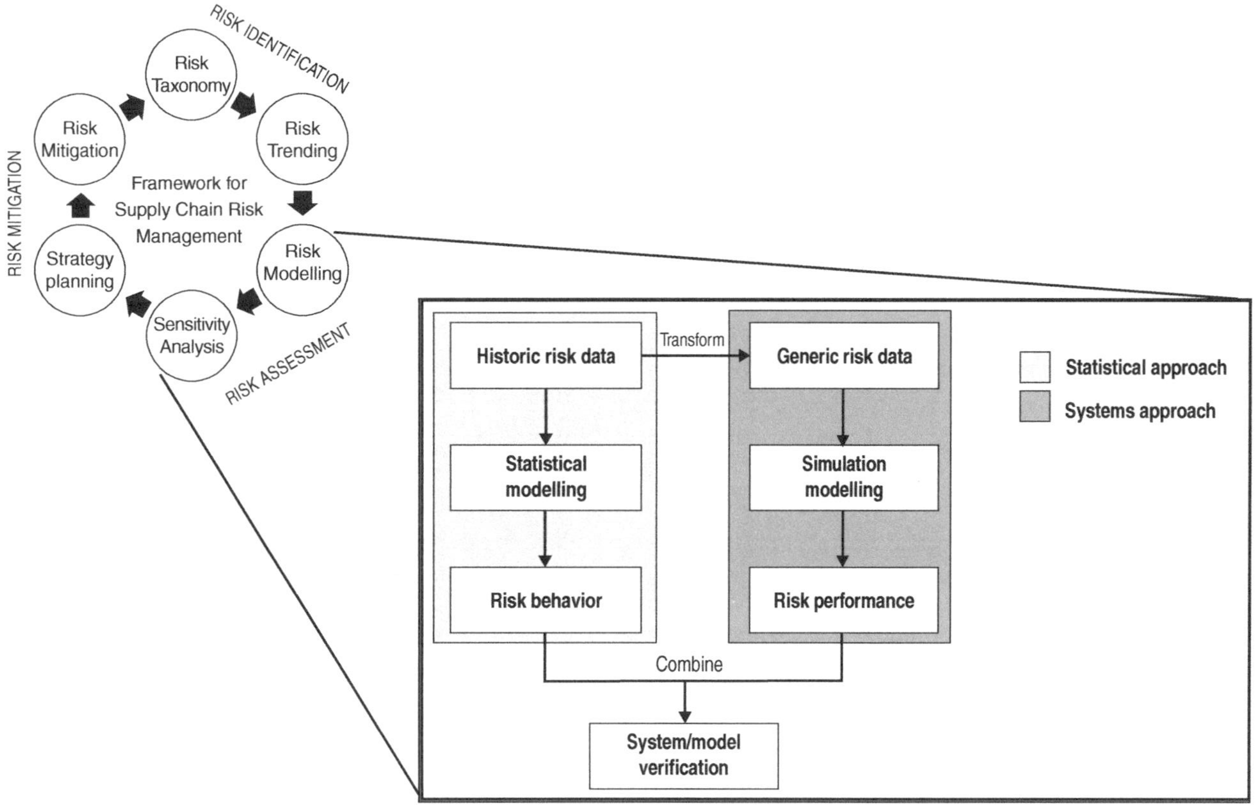

Figure 4.7 Research design: modeling supply chain risks.

transformed into a form required for the experimentation. The inputs from informal interviews with the Risk Managers were integrated to comprehend their understanding of possible risk impacts and severity of the events. The company reports helped in recording the events and their impact in terms of cost and delay over the running of the complete project. In order to comprehensively study the behavior of the risks, the available data was screened by filtering confidential information associated with the collaborating organization to form the historical risk data. In order to bridge the findings made from the qualitative and quantitative data sources, the Delphi method was used for arriving at a common consensus. The Delphi group, an extension of the focus group, is found to be a commonly used research method for data dissemination and learning. The Delphi method is used to obtain reliable consensus of opinion of group of experts with a controlled feedback system (McKenna, 1994). This structured technique is believed to work well when the objective is to improve the understanding of the problems and solutions (Skulmoski et al., 2007). Thus, the data available in different (qualitative and quantitative) forms was transformed into 'quantitative' historical risk data for experimentation. This transformed risk data comprised 30 risk events called 'risk scenarios', each having the description of event discussing type of risks observed and their probability, cost and delay changes over different stages/nodes in the project. The historical risk data was further transformed to form the generic risk data by sampling from the probability distributions. Historical risk data was analyzed following a statistical approach and generic risk data was analyzed using the SD modeling approach. Forrester advocated the use of computer simulation instead of mathematical models to learn about the systems modes of behavior and design policies to improve system performance (Vennix and Vennix, 1996; Lane, 2007). Richardson and Pugh III (1981) suggest that SD considers that 'feedback' and 'delay' cause system behavior and hence the system structure is very important to understand system behavior. Forrester (1961) suggested using SD simulation models for test-piloting a new structural form for an organization and investigating systemic challenges to supply chain network. The reason for the two modeling approaches used in this chapter was to test and validate statistical as well as empirical relationships between supply chain risks. Two distinct approaches were believed to facilitate critical insights through cross-comparison and combination of the results, difficult to comprehend individually. The SD simulation model for measuring overall risk performance is modeled using the simulation platform named Vensim®, which is a discrete event simulation software. The findings from the two different risk assessment approaches are collated and compared for drawing concluding results.

4.11 Modeling the Supply Chain Risks

The experimental study with the collaborating organization was conducted to test the viability of the framework in an industry environment. The collaborating organization that provided us with the data also provided the opportunity for testing the developed framework for SCRM. All stages from the framework are discussed systematically for predicting the overall behavior of risks within supply chain network.

4.11.1 Risk Taxonomy

Risk taxonomy can be defined as the method for facilitating the methodical and repeatable identification of risks associated within a given system (Carr et al., 1993). This particular activity needs to be comprehensive as well as consistent for the best process output. The first stage of the framework for SCRM is to identify and classify the risks based on causal (relational) attributes. There exist several risk classifications in the SCRM literature. Risk itself is termed as disruption, vulnerability, uncertainty, disaster, peril and hazard in the SCRM literature (Ghadge et al., 2012). A commonly preferred risk classification is based on 'sources of risk' as organizational and network risks. Organizational risks are the risks that lie inside the organizational boundaries, whereas network-related risks arise from interactions between organization and other supply chain network partners (Juttner et al., 2003).

The literature of ERM and systems thinking brings the concept of 'system of systems' where the enterprise or a larger system like supply chain is considered from a strategic (macro) as well as an operational (micro) perspective. In order to achieve this, we classified the risks based on multi-dimensional causal relationships seen in Table 4.5. This is not just limited to classifying the risks based on its risk sources but also takes into account other important inter-dependent factors such as work activities and business practices undertaken at an organization during the development of risk taxonomy. We adopted the *'enterprise architecture'* based classification from Burtonshaw-Gunn (2008) for identifying supply chain risks as this provides a systematic approach to selecting and recording unclassified behavior of risks. Enterprise architecture is classified into business and system architecture. The business architecture represents the most important work activities and assets in an organization along with the organizations core business practices as the primary set of requirements (Burtonshaw-Gunn, 2008). 'POLDAT' is the abbreviation for Process, Organization and Location, Data, Applications and Technology. POLDAT is a hexagonal model developed for process improvement and was first used by the American Computer Services Corporation for comparing the activities at different organizations (Burtonshaw-Gunn, 2008). The use of process improvement model for risk classification is expected to provide the systematic approach for capturing the risk behavior within the SC network. These six attributes are *'spheres of change'* which helps to identify commonalities between activities, issues, solution fits within a system (Burtonshaw-Gunn, 2008). These risk attributes constitute the portfolio of risks based on causality found in the enterprise or supply chain. It is essential to consider them together for a holistic picture of the risks within the supply chain network. This multi-dimensional perspective for classifying the risks utilizing theory from enterprise architecture is new to existing supply chain risk classifications discussed in the SCRM literature.

Table 4.5 Risk taxonomy: POLDAT (Adopted from POLDAT methodology, Burtonshaw-Gunn, 2008)

Enterprise architecture	*Risk attribute*	*Sources, activities, issues, practices*	*Nature of risks observed*
Business enterprise	Process	Focuses on the internal business activities. It looks at what the enterprise does and in which sequence it does it. Process attribute captures the end results by its classification.	Product design risk Information distortion risk Demand risk Quality risk Disruption risk Operational risks Financial risk,
	Organization	Focuses on human resources within an enterprise. It considers the culture, capabilities and roles of the people. It also considers the team structure and organizational units associated with the given activity.	Skill/performance risk Poor management Safety/Security risk Reputation risk
	Location	Focuses on geographic location types. Issues associated with physical and infrastructure facilities are considered in this set of attributes.	Supply risks Safety risk Geopolitical risk Supply risk Capacity risk
System enterprise	Data	Focuses on business information data. It addresses the content, structure and relationship associated with information data.	Intellectual Property (IP) risk Regulatory/Legal risk Information distortion risks
	Application	Focuses on structure, capabilities and user interface of the software used in the enterprise. All issues associated with IT are covered in this attribute.	Integration risk Network risk
	Technology	Focuses on hardware, technology associated with the software used. All issues associated with communication between hardware/software are considered in this attribute.	Technology risk IT failure

The risk registers for a project (new product development supply chain) were studied at the collaborating company. This led to the identification of 30 different risk scenarios and these were later classified by referring to their association with different sources, activities and practices within the organization. Based on this predefined risk taxonomy, some of the commonly observed risks identified from the risk scenarios are presented in Table 4.5. The nature of risks identified for each risk attribute is associated with either process or practice. This provides a good measure for not just classifying the risks but also provides a direct indication toward particular process needing attention to overcome impending disaster.

4.11.2 Risk Trending

It is necessary to understand the fundamental nature of risks before understanding the overall risk behavior. The risk attributes (POLDAT) are considered for the group of risks and then analyzed to draw a preliminary understanding of the risk profile. Risk trending is defined here as identifying (upper and lower limits) 'zones of operation' observed for each risk attribute. It is understood that every project or supply chain network is expected to behave independently and may have different operational limits. Risk is a financial liability (McCarthy, 1996) and hence it is important to define the limit of its liability. The operational limit also represents the worst-case scenario for driving insurance policies and project budgets. Some of the risk events comprised more than one type of risk attribute. In such cases, each risk was assumed to be independent with no appropriate distributions considered. Upper and lower limits of probability of event and its impact in terms of cost and delay are crucial parameters for the risk assessment process as they define the boundary of the system under study. Quality (of products and services), cost and delivery offered by the organization are the most important key performance indicators affecting the business performance (Ghobadian et al., 1994; Atkinson, 1999). At the same time, cost, customer responsiveness, quality and flexibility are most important supply chain modeling performance measures (Beamon, 1999). Quality and service associated with the customer responsiveness is assumed to be the function of either cost or delay (delivery time) in this risk assessment process. It is earlier identified by Gunasekaran et al. (2004) that the quality and service can be improved or controlled by additional cost or time.

Figure 4.8 depicts the static behavior of a group of risks classified using the POLDAT risk attribute taxonomy. The operating zones for different risk attributes are captured through a three-dimensional plot. The plot shows the operating envelope for average probability of the risk, impact cost and the duration of the risk for the analyzed data. The data available on risk scenarios was first collated into POLDAT risk attributes and later the average performance of probability, cost and duration was captured over different periods in a project. It is observed from the risk trending plot that the process-based risk tends to have a high probability at the beginning of the project compared with risks associated with location. The plot also gives information aboutthecostlimitsaswellasthegenericbehaviorofeachriskattributeoverthe length of the project. This kind of information could help Risk Managers to prepare proactively for the oncoming disruption. The generic static behavior provides first-hand information on the set of risks needing priority during the mitigation stage. The historical risk data was later studied to predict the probability distribution pattern of the risk performance variables. Different approaches for identifying the probability distribution are discussed in the academic literature. It is important to predict the right probability distribution fit for transforming the historic risk data into generic risk data for further quantitative analysis.

The scatter diagram shown in Figure 4.9 is a collection of points showing the relationship between dependent and independent variables. Using the identified

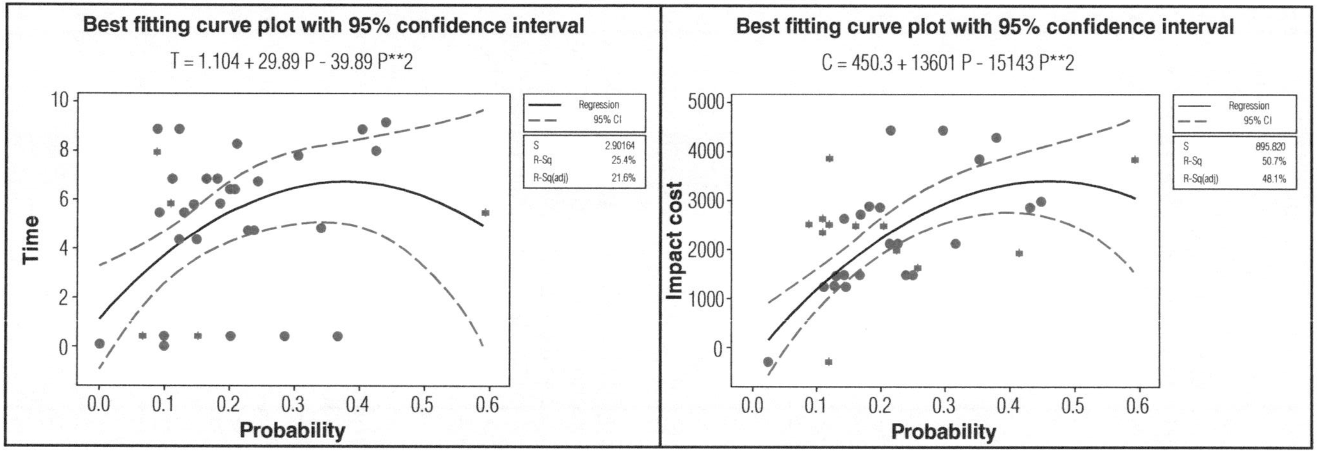

Figure 4.8 Risk trending: static behavior of risks (POLDAT).

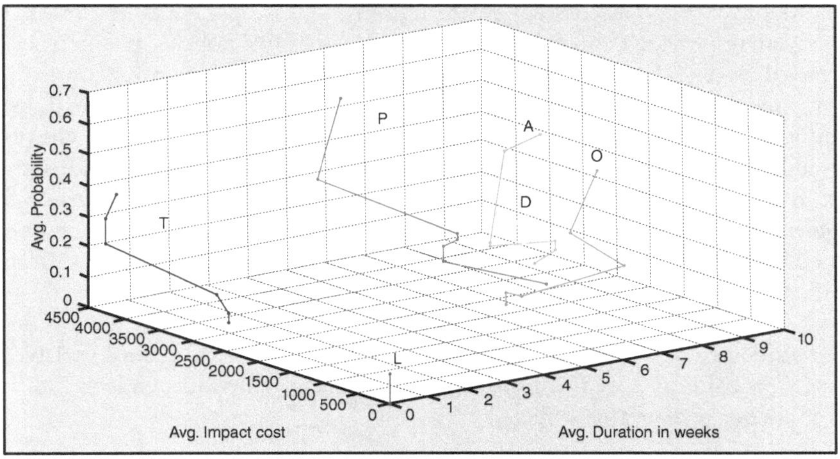

Figure 4.9 Best fitting curve for risk variables.

30 risk scenarios, three important risk performance variables, namely probability, cost and time (delay) are studied for any possible correlation. The behaviors of cost and time with respect to probability were plotted for all the risk scenarios. The scatter points obtained as seen in Figure 4.9 were analyzed for obtaining possible correlations between different risk performance variables. Minitab©, a commercial statistical and process management software, was used for generating the risk trending results. The best fitting curve attempts to obtain the possible 'degree of correlation', providing useful information for resources allocation during the project planning activity. Figure 4.4 shows the accumulation of risk scenarios in a specific range of probability, but does not provide evident correlation between performance variables. With a 95% confidence interval, the best curve fit for the set of data was found to be poor and hence had to be rejected. R^2, a 'coefficient of determination', is a statistical measure of how welltheregressionlineapproximatestherealdatapointsandisameasureofthe 'goodness of fit' for the estimated regression equation (Anderson et al., 2007). Lower values of R^2 were found as seen in Figure 4.4 for probability versus time and probability versus cost data points. No universal best-fit procedure is guaranteed to provide a correct solution for the random relationships (Ortells, 2011). This analysis is conducted with an intension to see if there is any significant correlation between the three risk performance variables. Probability, cost and time (delay) were observed to be behaving independently of each other for the given set of risk events. This means that even with the high probability of an event, there may be less likelihood of impact either on cost or time (delay) and vice-versa. With this crucial finding, further modeling of supply chain risks was developed.

One of the authors was closely associated with the organization and collected the qualitative data related to number of stages, their expected duration and risk operational limits for different past projects through informal interviews with Risk Managers. The discussions emerging out of the respondents in a research environment provide new concepts and critical issues like policies, competencies or causal factors (Luna-Reyes and Andersen, 2003). A focus group consisting of two researchers from SCRM and three practitioners from the Systems Engineering field formed the Delphi study group and provided consensus for the observed risk trending behavior. This activity was followed in three repetitions; the first focus group meeting did not derive any consensus but helped the group in problem synchronization. The second and third meetings led to a strong consensus on the relationship and static behavior of risk parameters. The focus group verified the assumptions made and supported in defining the boundaries of the system under study. Modeling of risks during the risk assessment process was later conducted with the hypothesis that the three risk performance variables, namely probability, cost and time (delay) are functionally independent and do not influence one another directly.

Notes

1 ***Upstream issues.*** Upstream risks are associated with procurement and are considered to be threats to supply assurance, the possibility of improper supplier selection, increased company liabilities and uncertainty in supply lead time (Smeltzer and Sifered, 1998; Sislian and Satir, 2000; Meixell and Gargeya, 2005). It is observed that about 56% of the related papers focus on upstream risks. The key issues of supply risks are found to be related to supply system design (number of suppliers (single/multiple sourcing)), location of suppliers (local/global sourcing) and supplier's agility, flexibility, delivery reliability and infrastructural strength and coordination and information sharing, which we covered in our classification. Analysis of the literature focusing on supply risks shows that information sharing and coordination issues have been paid the highest attention (44%) followed by the supply system design issues (36%).

2 ***Downstream issues.*** We focus on two discriminating elements and classify the demand issues as market volatility and demand fluctuation and coordination and information sharing. Coordination and information sharing among wholesalers, dealers and retailers and shorter planning horizons are some of the measures suggested in the literature to manage demand side risks (Gupta and Maranas, 2003; Chen and Lee, 2004; Boute et al., 2007; Stephen et al., 2007). There have also been proposals to investigate the level of information sharing from a security point of view and adopt trust-based mechanisms under volatile market conditions (Xiao et al., 2007). As mentioned in Table 4.7, issues related to demand and order variability have been considered more (63.5%) in the literature than coordination and information sharing issues (36.5%) to manage downstream risks.

5 Models of Risk in Supply Chain Management

5.1 Models of Risk in Supply Chain Management

The preliminary analysis on risk trending provided directions toward important considerations for modeling risks. The functioning of risk modeling is fundamentally based on a developed supply chain risk model. The developed model is a 'system' combining the risk theory and working mechanism for the risk modeling activity.

Figure 5.1 depicts the schematic of the supply chain risk model. 'Risk' is an input to the model taking into account different sets of risk attributes and parameters. The input requirements for the model to function are nature and combination of risk attribute; and the anticipated values of probability, cost and delay at the start of project. The model then considers the combination of risk attributes and their behavioral patterns to model the overall impact. The developed model considers a risk event triggered with an anticipated probability. For a given probability, it is expected to have a low or high impact on the supply chain system. Random integers are fed during this stage into the model to control the impact. The impact of the risk event could be high or low depending on the forces acting during risk propagation. This is presented in the model as high or low with a constraint that either one occurs during each risk event. In order to define the impact created by the risk event, a control feedback is provided which will calculate the impact just once (as high or low) depending on several parameters considered in the modeling. Although a risk event is assumed to be disrupting only once, in reality the risk impact propagates over periods and levels. Risk propagates in three different levels as primary, secondary and tertiary zone of risk propagation (Deep and Dani, 2009). In the primary zone of risk propagation, the disruption spreads into core activities within SC network, i.e. procurement, production and logistics. In the secondary zone, the risk affects critical service support such as R&D, Finance, Information technology and other non-critical supply chain entities. In the tertiary risk propagation zone, the risk further propagates to social and environmental elements of the business. Primary and secondary zones of risk propagation tend to have short-term impacts with tertiary zone having a long-term impact on complete supply network (Ghadge et al., 2011). Deep and Dani (2009) portray the primary, secondary and

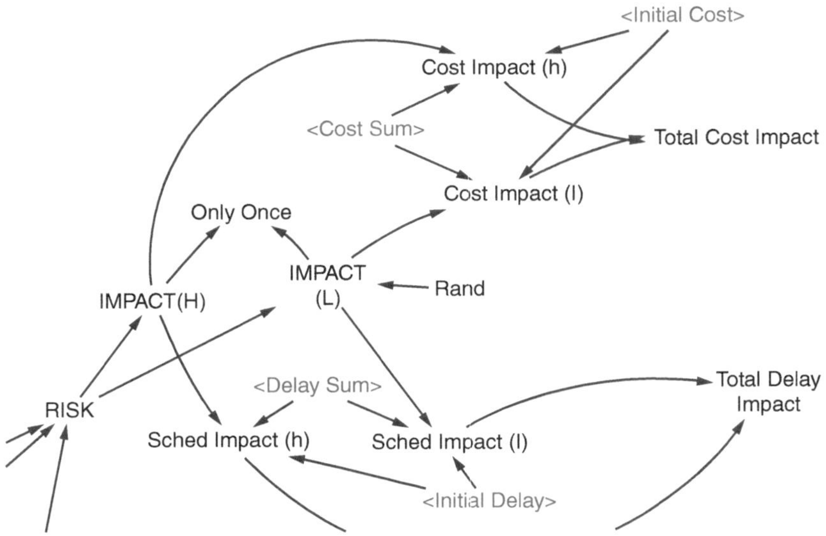

Figure 5.1 Supply chain risk model.

tertiary zones in a different way. They portray the primary zone as the critical chain of fulfillment; the secondary zone as the zone that feeds into the primary zone or is the output of the primary zone; and the tertiary zone as the zone that feeds into the secondary zone or is the output of the secondary zone. The developed model is designed to capture risk propagation phenomenon in periods within the primary zone.

A low or high impact condition for varying risk probability provides a condition for the risk to occur at a reduced impact providing early warning for disruption for possible mitigation action. At the high condition of impact, the probability reaches 100% (or more) and remains unchanged indicating the full extent of disruption, providing no opportunity for the risk mitigation. This concept of risk propagation is further expanded to capture the impact in terms of cost and schedule. The model later considers two scenarios for cost and schedule (as high and low). The accumulative impact in terms of cost and schedule over different periods is calculated as total cost impact and total delay impact, respectively. The cost and time (delay) impacts are associated with the overall impact of disruption and hence separated for individual assessments in the model.

For the smooth functioning of the model, projected or anticipated values for initial probability, initial cost and initial time (delay) were provided to activate the system. The system model automatically considers the previous parameters for measuring the impact for the next period. The overall cost and time (delay) accumulated over the period were represented as total cost and delay impact. The risk performance was evaluated in the form of impact for the given probability.

Based on this underpinning concept, statistical modeling was performed to predict risk behavior, whereas the simulation modeling was performed to predict risk performance.

5.2 Management and risk

In the present chapter, the history of development interrelation between theories of management and risk is stated. Causes and consequences of large catastrophes and accidents are considered: the most dangerous manufactures are indicated, and risk values and possible damages are shown. A classification of sources of catastrophes and accidents is given. Two different approaches to risk management on the basis of active actions and insurance are considered; the role and place of monitoring in risk management are discussed. General theses of the State Safety Program of Russia are presented. The role and place of the nonlinear mechanics methods, of the theory of probabilities and of the logic and probabilistic risk theory in modeling and risk management of catastrophes, non-success and accident are considered.

5.3 History of Interrelation of Management and Risk

Management and risk existed at all times from the moment of appearance of mankind. Management provided existence of each human being and the whole human community. First, the management was empirical; it was performed with account of risk on the basis of intuition, experience and common sense. At later stages of mankind history, the states appeared. Then, management was performed by the Supreme governor of the country on the basis of the code of rules and directives of religion. The basis of such management keeps both in society and engineering up to our days. Later, for more efficient management, the elements of the mathematical management theory and the mathematical optimization theory began to be used in practical resolving of particular problems.

During Industrial Revolution, the classical theory of management (regulation) of separate mechanisms, devices and processes, based on the description of dynamics of objects in terms of differential equations, was created. In management, the risk was taken into account indirectly by using criteria of stability, opportunity of the resonant phenomena, destruction, etc. Successes of the classical theory of management are enormous; as an example, management of start and movement of a spacecraft should be mentioned. Amidst the main contributors to the classical theory of management are H. Chestnut, R. W. Mayer, F. R. Bellman, L. S. Pontryagin, J. Z. Tsypkin, etc.

During the Second World War, purposes of management stimulated formation of mathematical disciplines such as operations research (John von Neumann, etc.); the theory uses the system approach to statement of tasks and decision making. Later, this discipline switched almost completely to the theory of games and the resolving of optimization tasks by methods of linear and nonlinear programming; methods for resolving of separate tasks of optimization with criteria of economic efficiency (transport problem, cutting materials, etc.) were created.

Immediately after the Second World War, Norbert Wiener formulated principles of the cybernetic control theory. In the theory, observable input and output parameters of an object are used to create the mathematical model of the object, named *"black box"*. Such management was used for resolving particular problems of optimal control. The risk with such management was considered as the probability of failure in achievement of the purpose due to inadequacy of the model and the presence of hindrances.

In 1952, the theory of management of risk of investments appeared, when H. Markowitz formulated the problem of choice of an optimal security portfolio. In H. Markowitz's consideration, yield as the mean value and risk, as the mean square deviation and as the measure of uncertainty of yield was taken into account for each security in a portfolio. Such new concepts as diversification, Indifference Curves of the investor, achievable and effective sets of portfolios were introduced. The significant contribution by H. Markowitz was marked by the Nobel Prize in economics in 1990. Further, the portfolio theory was developed by D. Tjubin, D. Marshall, W. Sharpe and S. Ross who were also awarded by Nobel Prizes (Sharp et al., 2001).

Computers' coming into the being allowed V. Glushkov, V. Skurihin, etc., to create the technology of information management, namely the automated control systems (ACS) (Skurihin and Morosov, 1976; Glushkov et al., 1989). These systems have well-structured database, information technology with the window interface, software for resolving the certain type of optimization problems, expert systems for decision making, software for forming reports and printing of illustrations. The systems allow to give out any information at inquiry or to resolve problems, to find areas of optimal admitted decisions, to choose the most effective solutions. Acceptance of the final unique decision is last to expert. Within the framework of ACS, the problems of numerical risk estimation were not resolved.

New step in the development of the management theory was the formation of situational management on the basis of logical-linguistic models (Pospelov, 1976). It was shown that the management of complex objects is impossible in Principe without taking into account the qualitative semantic information which cannot be expressed quantitatively. For the first time in the theory and practice of management, the logic, sets and logic connections of objects and events were introduced. Various approaches were suggested for the description of observable situations, based on languages with advanced semantics; various methods of construction of knowledge models were presented, allowing to reflect in the model's qualitative proportions and the rules inherent to the object; various procedures were given for finding solutions to problems of management, based on logical-linguistic models. The considered theoretical results find applications in problems of operatively dispatching control in seaports, airports, etc. Problems of risk in systems of situational management were not studied. Further, this concept was developed in Ivanishev and Marley (2000).

Of great importance was the formulation of logical-probabilistic methods (Ryabinin,1976, 2000) for quantitative modeling and analysis of reliability and safety of structural complex technical systems. These logical probabilistic methods (LPM) are a special section of mathematics connected to the logical-probabilistic calculus. These methods make it possible to sort elements of complex system (CS) according to their importance. These methods have passed approbation in real projects of Navy fleet. They have become the intellectual core of control systems of reliability and safety in many complex technical systems.

Further development of logical-probabilistic methods was done in methodology of automatized structure and logical modeling (Mojaev and Gromov, 2000). The methodology makes it possible to use all logical connections (AND, OR, NOT) and introduces schemes of the functional integrity. The latter allows one to represent the scenario of successful or unsuccessful functioning of technical or organizational system as a graph including fictitious nodes. The software for numerical structural and logical analysis of stability and efficiency were developed. They were successfully used for the educational purposes and for resolving various applied problems of the analysis and management on the bases of suggested methodology.

On the basis of logical-probabilistic approach, the theory of LP-modeling and analysis of risk with groups of incompatible events (GIE) was created (Solojentsev et al., 1999; Solojentsev and Karassev, 2003). The theory made it possible to model and analyze risk in systems, where elements and the system itself have some possible conditions, and to apply the LP-models with GIE for quantitative modeling and risk analysis not only in technical, but also in economical, organizational, ecological systems. Conditions of elements in systems could be described both quantitatively and qualitatively. High accuracy and robustness of LP risk models are stipulated by the use of Bayes' formula and by the generation of well-organized probabilistic risk polynomial. LP risk models with GIE use discrete nonparametric distributions of probabilities of grade-events in GIE. The latter makes possible calculation with multivariate distributions; the distributions are arbitrary and can have "heavy tails". In LP risk models with GIE, dependence and coherence of variables are taken into account on the basis of combinatorial theory ("everyone with everyone") and the theory of correlation is not used. The LP risk model with GIE allows one to perform active scenario risk management instead of passive risk management of insurance in other methods. Means and the maintenance of scenario risk managements of failures and accidents in CSs on design stages, and in testing and operation are described on the basis of the LP-theory of risk with GIE.

At the present time, two different components in the theory of risk management on the basis of active operations and passive insurance are intensively developed and their optimal combination (Pechenin, 2000) is sought. Here, the great achievements in development and use of monitoring of risk management in business and engineering, which allows one to make decision with open eyes (Krasnov, 2002), should be mentioned.

Besides, of interest are studies of scientific bases of information (Ivchenko et al., 1997; Yusupov and Zabolotsky, 2002), where the conceptual bases of the information theory are considered, its essence, purposes and principles are determined and formulated, problems of information theory and ways of their resolving are shown, the basic stages and directions of development of information theory are determined, dual nature of mutual relationship between science and information theory is revealed; inevitably, the problems of informational safety are considered too.

In Russia, works on strategy of risk management with application of new approaches from the area of fundamental sciences started in 1997. In the book *Risk Management* (*Risk Management*: 2000) by famous scientists, who are also the authors of the State Program "Safety of Russia", special attention was paid to problems of strategy of risk management. The concept of authors is the assumption that the mathematical theory of safety and risk can be constructed on the basis of the accumulated experience of the new science. This theory would take place between the level, where political and strategic decisions such as laws are made, and the level of development of concrete technical systems. As a methodical basis for creation of such theory, the use of nonlinear dynamics was suggested. We note that the latter point can be true only for accidents such as earthquake, floods, snow avalanche, etc., characterized by slow accumulation of energy or weights with their further very fast freeing up.

In most cases, accident in human–machine systems occurs when some events happen simultaneously or the risk of condition of system and its elements as a result of "deterioration" exceeds the admitted value. Even, an example of a human being clearly shows that the person becomes tired, and requires rest and food in order to prevent him/her or a technical system, which he/she controls, from accidents. Here, another approach is necessary to model the risk of failures and the accidents, which would be alternative to methods of the nonlinear mechanics. We shall name such approach logical-probabilistic or scenario approach for the management of risk of non-success.

5.4 Monitoring and Risk

Monitoring is the integral part of safety security in technical, economical, organizational and social systems. An example of monitoring is given by the world economics. Really, a large number of daily and weekly economic newspapers inform us about costs or stock indexes of companies, about exchange rates, sales volumes, etc. There are numerous independent institutions and agencies which estimate and publish ranking of banks, countries and branches, the reliability of capital investments, etc.

Now by using Internet, it is possible to follow in the real time (with a delay of minutes) the situation on all main financial and commodity exchanges of the world in New York, London, Chicago, Tokyo, etc., including sales volumes, a pent-up demand, exchange rates, indexes of stocks, the prices for grain, cotton, petroleum, gas, gold, copper and other metals and the goods. The same detailed information can be obtained for any period in past on minutes, hours, days, months and years. Everything in business is made with open eyes. The openness of information is the reason why the world economics for the last 70 years has not been in such sharp crises, as in 1929.

Monitoring of such kinds of sports as chess and tennis allows sport organizations to rank players according to their results and, thus, to solve the problem of formation of the lists of participants and optimal scheduling of tournaments.

Monitoring in medicine based on patients' disease records, including their cardiograms and analysis data allows physician to organize effective and safe treatment.

Monitoring of the society state via public-opinion polls on various subjects makes it possible to reveal the most urgent problems of the society to prevent social explosions and to plan effective programs of development and reforms.

For complex technical systems, buildings and constructions, intended for long-time operation, failures and accidents can be caused by degradation of properties of materials, by reaching limit levels of the accumulated damages, by formation and uncontrollable propagation of cracks, by cavitation wear, by breakdown of tightness of flanges, by reduction of resistance of isolation of cables due to aging polymeric coverings, etc. For potentially dangerous objects and manufactures, the essential exhaustion of the design resource is characteristic. In crucial branches (power, petrol and chemical plants), potentially dangerous objects have exhaustion of designed resource at the level of 75–90% (*Risk Management*, 2000).

5.5 Scenario of LP-Modeling and Management of Non-Success Risk

In CSs, the scenario of failures and accidents has logical and probabilistic nature. Therefore, we write a scenario of non-success or catastrophe and the logic function of risk and build the probabilistic polynomial of risk (Ryabinin,2000). Elements of CS have logical connections OR, AND, NOT and cycles. The mathematic basis for the estimation and analysis of risk is the logic and probabilistic calculus. The analysis of the risk allows us to manage risk.

Probabilities of failures of elements can change in the course of time (the elements wear out, age, collapse, depreciate, etc.). The risk LP-models with dynamic interpretation are much more constructive and clearer than with differential equations, so they are true with high probability.

Each CS has some value of risk of safe functioning (and each element of the system has some value of risk of non-success or failure). If the risk becomes more than the admitted one, the system either cannot be supported by itself, or it is useless, or harmful. Then, the system ceases the existence (for example, a bank bankrupt). Or there occurs a serious structural reorganization of CS when some elements disappear and some new ones are brought to the system. The latter changes logical connections and probabilities of failure, so a new CS appears.

Thus, it is possible to simulate visually all past and present catastrophes on the basis of the LP-approach without using the mathematical apparatus of the nonlinear mechanics (*Risk Management*, 2000) and the classical theory of probabilities. Many examples of real catastrophes can be interpreted by LP risk models (Frolov and Mahutov, 1992; Problems of destruction, 1997) with varied probabilities of elementary events and existing admitted risk for a CS in appropriate environment.

The scientific basis of the risk LP-theory and technology of the scenario logic and probabilistic management by risk are: the LP-calculus, the LP-methods, the LP-theory for systems with GIE, the theory by Markowitz and the theory of VaR for risk of security portfolio. Besides, it implies using discrete mathematics, combinatorial theory, Weil's theorem, nonlinear optimization and modeling of Monte-Carlo and algorithmic calculations on modern computers.

5.6 Reasons and Consequences of Large Accidents

Development of the environment created by technologic activity of mankind in XX century occurred at much higher rates, than in previous centuries. It has resulted in two opposite consequences both in industrial countries and in the rest of the world (Accidents and catastrophe, 1995–1997; *Risk Management*, 2000):

- Outstanding results in electronic and nuclear industry, airspace, power and chemical engineering, in biology and gene engineering, which advanced mankind to essentially new boundaries in all areas of activity, were achieved;
- Unprecedented earlier potential and actual threats to a human being, to objects created by people, to local and global environment acting, not only in military, but also in a peace time, were created.

Thus, the center of attention moved from dangers to risks – from mudflows, typhoons, flooding, earthquakes and other natural phenomena, to man – caused, ecological, social disasters, stipulated by decisions, accepted by people.

For the first time, the special attention of the public and scientists to large industrial failures was attracted after disasters in 70–80s of XX century at the chemical enterprises in Flixborough (England, 1974) and Seveso (Italy, 1976); then, as a result, hundreds of people were affected, there was essential, irreparable damage to environment and huge resources (material, human, time, etc.) were spent for liquidation of their consequences. In 1980s, the tragedy in Bhopal (India, 1984) and Chernobyl (Ukraine, 1986)) perpetual virus attacks in the Internet, and large-scale acts of terrorism in USA (September, 2001) continued the list. As a result of accidents, enormous damage to environment was caused, and the amount of lost people was measured by thousands (Zhelesnuakov, 1998).

Strengthening of two types of dangers (Frolov and Mahutov, 1992; Problems of destruction, 1997) is observed in natural and technogenic spheres. First, it is the well-recognized ecological dangers for nature, as the living environment, caused by persistent negative anthropogenic pressure on environment. Increase in these influences in combination with global natural processes of change of climate and environment can result in ecological disasters of global and national scales. Second, the rapid scientific and technical development in civil and defensive areas in many countries of the world has resulted in essential gap between exponentially growing threats in natural and technogenic spheres and ability of each country and the whole world community to withstand these threats.

The level of a person's safety, of safety of states and the whole mankind, of the natural environment from all increasing dangers of natural and technogenic accidents does not raise yet despite the efforts undertaken everywhere in the world. It is notable that natural and technogenic accidents are capable to create and strengthen threats in sociopolitical, economic, demographic and strategic spheres.

The insufficient ensuring of safety results in annual losses, measured by billions of Euros. Problems of safety and risk in ecology, engineering, finance and economics, terrorist and information danger have become actual problems of state scale.

Today in Russia, there are about 45,000 dangerous manufactures, a great number of constructions, whose destruction can result in disasters not only of regional, but also of national scale.

Many countries, including Russia, are facing the necessity of liquidation in the shortest possible time of large-scale extreme situations (ESs) having non-military character. If the ES arises in industrial area, or large city, it inevitably causes significant destructions and losses, and hundreds and thousands of human beings can be lost.

A great number of ESs happen annually in the world. In 1994 in the Russian Federation, 1,076 technogenic ESs occurred. Much of ESs happen in industrialized territories. A number of technogenic ESs essentially increased in Northwest (91%), Central (48%) and Transbaikalian (41%) regions.

5.7 The Most Dangerous Industry Branches

According to the level of potential danger resulting in accidents in technogenic civil sphere, it is possible to give extra attention to objects of the nuclear, chemical, metallurgical and mining industry, unique unusually large-scale engineering constructions (dams, viaducts, oil storages), transport systems (space, water and underwater, ground), which carry dangerous cargoes, and a large number of people, gas and oil pipelines. Many military objects such as space-rocket and aviation systems with nuclear and traditional charges, nuclear submarines, and large warehouses of usual and chemical weapons should be mentioned too.

5.8 Values of Risk and Damage

For providing the technogenic safety on the boundary of the XX and XXI centuries, it should be taken into account (*Risk Management*, 2000) that in global technogenic environment, both in civil and military spheres, there are about 10^3 objects of nuclear engineering for peace and military purpose, more than $5*10^4$ nuclear ammunitions, about $8*10^4$ tons of chemical armament of the mass destruction, 100,000 tons dangerous explosives and strongly acting poisonous substances, and 10,000 objects with high reserves of potential and kinetic energy of gases and liquids.

In the analysis of safety of technogenic sphere along with the aforementioned damages, it should be taken into account whether the corresponding potentially dangerous objects are made in series. The heaviest accidents are characteristic of unique objects, i.e. produced in the single copy or in small series. The number of nuclear power reactors of the same type is 1–10 with their general number 450–500 in operation, and the number of the same space-rocket systems is from 3–5 to 50–80. Medium-series potentially dangerous objects are estimated by hundreds and thousands, and large-series are made in tens and hundreds and thousands (cars, agricultural machines, machine tools). In connection with the aforementioned, the integrated economic risks, which are determined by multiplication of individual risks by the number of objects, are comparable for accidents of big objects and for accidents of many small objects.

Of high importance is the level of substantiation of safety of potentially dangerous objects achieved in designing. With reference to failures of large-series complex technical systems, where dangerous damages arise in usual conditions of operation, the level of forecasting of safety and reliability is 10–100%. Dangerous and catastrophic destructions of large- and medium-series complex technical systems in conditions of normal operation are predicted in much smaller measure – from 1% to 10%.

From information about probabilities and risks of technogenic failures and accidents on objects with extremely high potential danger, it follows that the difference in the levels of required and admitted risks, from one side, and the level of realized risks, from other side, reaches two and more orders. At the same time, it is known that increase of the level of security of objects from failures and accidents by one order only requires huge efforts in scientific and technical spheres and the expenses being comparable with 10–20% of the project cost.

"Generally, as CSs, we shall..." understand the structural complex human–machine systems consisting of the equipment, computers, software and actions of the personnel both having elements and output with several conditions. The appearance of emergencies, failures and accidents in CSs such as nuclear power plants, starting rocket systems, oil- and gas processing and other chemical manufactures, pipelines and transport systems, is usually classified as rare casual events. However, in view of the consequences such as emission of radioactive and toxic substances, explosions with scattering parts of construction, extensive fronts of flame and pollution to the environment, the biggest of the disasters can be compared with large-scale natural ones.

The reasons of failures and accidents in CS, depending on their developers, manufacturers and consumers, are:

- insufficient quality of projects;
- insufficient quality of development tests;
- insufficient quality of operational tests;
- insufficient quality of operation monitoring;

- deterioration and aging of the equipment in operation;
- decrease of quality of the work of personnel due to influence of social
- factors;
- mistakes of the personnel;
- swindle of the personnel;
- terrorist actions;
- attacks of hackers.

Actions of these reasons both separately and in their combination result in failures and accidents with human losses (both personnel and the population of region), with large material damage, danger for the environment and decrease of living standard of the population.

We note that both experts and the public are paid insufficient attention to some of the mentioned reasons of failures and accidents, because of their appearance with delay; the latter explain the absence of interest of developers in spending extra money to the project safety and the tendency of owners to hide true reasons of failures, unsatisfactory quality of testing of systems. We mention such underestimate reasons as an example.

6 Risks in Environmental Management

6.1 Risk Management and Insurance

We consider features of risk management using a historical example (Pechenin, 2000) of approaches to the estimation of danger of sea pirates' attacks, the so-called *Bernoulli's* and *Columbus'* approaches. Two-hundred and fifty years ago, *Bernoulli* found a way to reduce the insurance tariff at insurance of merchant. Using low tariff, he drew the clients, and due to the big number of clients, he could achieve sufficient accuracy in the calculation of probability of loss of the goods or the vessel, and with the low insurance tariff, he could get a good profit.

Two-hundred and fifty years earlier, *Columbus* started searching a way to India. For his ships, as well as for the merchant ships of *Bernoulli's* time, the main threat was the pirates. The probability of attack of pirates was high, but whether it was necessary for *Columbus* to know the value of this probability? *Columbus* equipped the ships with rectangular sails of the maximal area. He lost the maneuverability, but this essentially increased the speed of caravan. On the second day of expedition, a pirate-sailing vessel approached *Columbus'* ships; however, some days later, it lagged behind hopelessly. It is necessary to notice that the pirate ships had greater maneuverability, than the trading ones, and high speed. But their sails were universal, adapted to fight maneuver, and had no such large area as sails of Columbus' ships.

The given facts from history illustrate two approaches to the risk estimation. First approach *(Bernoulli)* assumes that process, which failure risk is necessary to estimate, cannot be adapted or it is not controlled consciously. Second approach *(Columbus)* is applicable to processes to which failure risk should be reduced ad infinitum by appropriate adjustment.

Bernoulli's approach does not demand an investment of money and efforts to the transformation of process, to which failure risk is estimated. It is the passive financial approach. Permanent updating occurs because a new process is generated instead of unsuccessful process. The approach is applicable to processes, where the failure costs are lower than those of the process adjustment.

Columbus' approach, on the contrary, should be applied to processes, where failure costs appreciably exceed the process adjustment costs. This approach is troublesome, but expenses for its realization grow linearly depending on complexity and danger of process, and costs from failure of complex and dangerous processes grow till geometrical progression. Thus, with some complexity and danger of process, the approach of Columbus appears to be economically reasonable.

Nuclear insurance pool successfully illustrates absurdness of *Bernoulli's* approach to the insurance of nuclear and radioactive dangerous objects: even for 100 years, it is impossible to generate the pool, sufficient for liquidation of consequences of failure of Chernobyl's type, as the enterprises are not able to pay insurance tariffs.

The aspiration of the insurance company to be prepared for the failure of Chernobyl's type is nothing but an attempt to resolve the ***Columbus'*** problem by ***Bernoulli's*** methods. ***Bernoulli's*** approach is applicable in its original form, if:

- insurance cases come rarely, values of insurance premiums are big enough, but insurance tariffs for the large number of the same objects of insurance cover costs of the insurance company, which can work for long and effectively;
- insurance cases come frequently, values of insurance premiums are not significant, insurance tariffs do not constrain economically the activity of the insured enterprises and cover costs of the insurance company, which can work effectively;
- insurance cases are coming with any period but the value of insurance premiums changes over a wide range and from time to time can put the insurance company on the face of the crash. In this situation, work of the insurance company in ***Bernoulli's*** approach assumes inevitable bankruptcy when the most serious insurance cases occur.

Application of ***Columbus'*** approach in the insurance of dangerous and expensive objects eliminates the possibility of the appearance of failures such as Chernobyl.

6.2 State Safety Program of Russia

As a rule, failures and accidents are followed in a short time by a flash of activity of "government officials" on the creation of the commissions for the investigation and distribution of welfare payments. Charges of the Ministry on Extreme Situations are going to take soon a quarter of the budget of the country because of the increased number of failures and accidents.

Their "work" on overcoming consequences is visible "on the face". To ensure the work on the decrease of a risk level of failures and accidents is much more difficult, as it needs new approaches, strategy, principles and methods, new culture and means. Results of these efforts will be visible only in some years or remain unnoticed if serious failures and accidents do not occur.

The analysis of results of examinations of large man-caused failures and accidents of XX century shows that the further development and realization of programs of scientific and technical development of the modern civilization and operation of CS is impossible without the system scientific approach to solve the problem of maintenance of safe functioning of similar objects and development of the methodical apparatus for quantitative risk estimation.

Creation of fundamental scientific, lawful and economic bases of providing the safety is one of the purposes of the state scientific and technical policy and the state scientific and technical program on the safety of natural and technogenic spheres, intended to increase safety in the industrial, energy, transport, building, oil-and-gas, mining and defensive branches, in production of new materials and technologies.

The state program "Safety" defines and fixes transition to the analysis and management of risks, as the basic system of regulation and safety, instead of the existing approach of maintenance of absolute safety [see (Frolov and Mahutov, 1992; Kusimin et al., 1997; *Problems of destruction*, 1997; *Russia Safety*, 1998, 1999; *Risk Management*, 2000; Vladimirov et al., 2000; Frolov and Bulatov, 2001; Mahutov, 2001; Mahutov et al., 2001)]. The state strategy is intended to provide formation, acceptance and use of scientifically approved methods and criteria of definition and management of conditions of systems in the parameter space of admitted risks.

The purposes of the state strategy are as follows:

- Controllable and normalized state, regional, branch and object management of creation and functioning CS by new risk criteria.
- Optimization of actions in extreme situations for the minimization of their immediate and distant consequences.

Ways for reduction of risk and softening of consequences of extreme situations follow from the general principles of safety in natural and technogenic sphere: priority of safety, high level of state regulation, use of risk analysis methods, inevitability of the personal responsibility, obligatory compensation of damage, availability of information, declarative way of activity, analysis of extreme situations, etc.

The fundamental problem of modeling and analysis of safety of CS includes tasks: creation of scenarios of failures and accidents and construction of mathematical risk models, development of methods for providing safety of operator, working personnel and population in case of emergencies in CS.

6.3 Methods of Nonlinear Mechanics and Probability Theory for Accidents

Nonlinear mechanics methods. In the state program "Safety of Russia", hopes are laid for use of methods of nonlinear mechanics for forecasting and modeling of accidents (*Risk Management*, 2000). For that, a number of possible approaches and models are considered: regimes with intensive development as analogues of the catastrophic phenomena, strong turbulence as a mechanism of origin of the accidents, the self-organized criticality as the universal mechanism of accidents, the theory about channels and jokers, etc. Since the formulation of these approaches in the program is rather declarative, we shall describe these approaches to estimate their applicability for modeling and forecasting of accidents.

Regimes with intensive development as analogues of the catastrophic phenomena. In order to forecast the catastrophic phenomena in complex organized systems, it is necessary to answer a number of key questions:

- whether the structure of system permits phenomena of this kind?
- in what elements (areas) an accident can happen?
- when it will take place; whether it is possible to estimate the time of development of accident?
- what part of structure of the system is determining for the occurrence of accident?
- of what kind and how heavy could damage be?
- how should the structure of the system be changed or how should the governing influences be regulated to prevent the accident?

It turns out that these questions stated in different terms raise and are answered in the theory of the nonlinear parabolic differential equations (*Risk Management*, 2000). The parabolic equations make a basis of the mathematical models describing the evolution of various processes in physical, chemical, biological, social, economical systems. For example, the equations are used in the theory of nonlinear heat conductivity, diffusion of the charged particles in plasma, filtration of gases and liquids in porous environments, in chemical kinematics, in problems of description of evolution of populations.

Strong turbulence as the mechanism of origin of accidents. It is suggested to develop an approach to investigate the problem of occurrence of rare catastrophic events in systems with complex behavior on the basis of synergetic principles. The essence of the approach is the observation that fast processes in systems are often determining, or at least very essential and seriously influencing behavior of slow processes. Therefore, of most interest is the charge, which spontaneously arises in the system and quickly develops in large scales. An example of such process is the development of a crack in a solid body. However, this example is not interesting, because as a result of the process, the initial system disappears. Another example is the appearance of large hurricanes and typhoons in the system "atmosphere-ocean". In this case, "large-scale perturbation" does not result in the destruction of the system, but leaves an essential trace.

Nevertheless, it is difficult to create models for the phenomena of this type. In view of the events, being rare, the conditions of their rise, even if they are determined, are usually treated as the extremely rare combination of improbable events. In the models, it is very difficult to find the key factors and parameters of order. Among the models with exponential distribution, the model of "heap of sand" is most popular in the theory of self-organized criticality. As events, the theory considers massive avalanching from the heap on which separate grain of sand falls. Such models are described using a strong turbulence concept in the Ginzburg–Landau equation.

The self-organized criticality as the universal mechanism of accidents. Here, an event is treated as catastrophic or dangerous, if it appears unexpectedly (i.e. it cannot be predicted) or if it is extraordinary (i.e. it is distinguished from a set of events, related to it), or both. In either case, it is possible to conclude that the system inducing this event is a complex system, because from simple systems, it would be natural to expect a clarity and predictability, from one side, and uniform behavior, from another side.

Though rigorous definition of concept of complexity does not exist, experience by the development of synergetic and studying of real systems intuitively determined as complex allows us to state some common ideas about properties of any complex system at different levels of the description.

1 At the mathematical level, complexity is intricacy related to nonlinearity of the description, since for linear systems, we apply the principle of superposition, allowing us to consider independently various working factors, parts of system, etc., that guarantees its simplicity.
2 At the physical level, the description, as a rule, is possible only in statistical terms, such as the density of probability, correlation, the mean of distribution, dispersion, etc. It occurs either due to chaotic behavior, specific for many nonlinear systems, which limits possibilities of the determined description, or in view of a very large number of elements, combining the system, what makes such description practically useless.
3 At the philosophical level, the following observation is essential: the more sophisticated and specific mechanism of some phenomenon, the less often it should be realized. Besides since virtually anything in nature is somehow connected to complexity, the mechanisms, laying in its basis, should be simple and universal.

From the above-stated, it follows that the investigation should be concentrated on the universal nonlinear mechanisms, resulting in complex behavior, demanding statistical description. Thus, in the study, it is possible "to find a back door" to generalize the data on the investigated complex systems and on the base of this material to try to give the description of the mechanisms, laying on their basis. Below, we shall consider the manifestation of complexity, and the theory of the self-organized criticality. Besides, we shall give a review of some self-organized critical models.

Discussion. It is typical [see (*Risk Management*, 2000)] that despite loud promises and declarations of applicability of the above-mentioned and other nonlinear methods of mechanics in physical, chemical, biological, social, etc., no concrete example of modeling of real accident is given. It is easy to explain because it is impossible to write down the differential equations for laws of conservation of energy, mass and amounts of movement for complex systems; it can be made only for the simplest systems and elements.

As early as 50 years ago, outstanding scientists John von Neumann and Norbert Wiener wrote about the impossibility to write down the differential equations describing behavior of complex systems. They stated that mathematical methods, which would be developed for complex systems, would be based on logic, combinatorial theory and the set theory, but not on the differential equations.

Let us remind the rule "Occam's Razor" (Guding et al., 2001) which is applied when in conditions of uncertainty or incomplete information; for the description of complex natural and social phenomena, it is necessary to choose one of two or several theories (methods). The meaning of this rule is that simpler explanations of the phenomena have a high probability to appear correct, than more complicated ones. In other words, it is reasonable to choose the theory, which includes the least possible number of assumptions or involved argumentation. The sense of the metaphor, giving the name of the rule in question, is in cutting off the superfluous principles and reduction of model to possible minimal number of assumptions. "Occam's Razor" is an extremely useful but seldom used methodological tool.

We omit discussing other nonlinear and logical and probabilistic methods which have proved to be successful. We only note that all the methods meet demands of "Occam's Razor" principle and recommendations of John von Neumann and Norbert Wiener.

Probability theory methods. For modeling of risk, it is proposed to use the informational-statistical approach for the formation of risk models and their identification from limited information on the basis of analytical laws of distribution of random values (Han and Shapiro, 1969; Ivchenko et al., 1997; Ivchenko and Martishenko, 1998, 2000). There are a lot of different distributions, but they cannot rigorously and precisely predict rare events of the real world, and it is proposed to improve these distributions by considering parameters of distributions as random values. At this, methods of randomization of Poisson parameter, generating functions, principle of maximum uncertainty and Lagrange's probabilistic distribution are used.

In this way, the following distributions are obtained:

1 The Poisson distribution, where ν is a quasi-determined value.
2 The modified Poisson distribution, where ν is distributed by the normal law with known parameters.
3 The modified Poisson distribution, where ν is distributed by the normal law and estimations of parameters m_ν, S_ν^2 are known.
4 The modified Poisson distribution, where ν is uniformly distributed over a known interval.
5 The Pascal distribution (negative binomial distribution), where the law of distribution ν is approximated by the gamma distribution with the form parameter m and the scale parameter λ.
6 The non-central negative binomial distribution, where the law of distribution ν is approximated by the gamma distribution with the form parameter m and the scale parameter λ.
7 The Poisson distribution of the degree k, where the law of distribution ν is approximated by the gamma distribution with the form parameter m and the scale parameter λ.

8 The beta geometrical distribution of the degree k, where the law of distribution ν is approximated by the gamma distribution with the form parameter m and the scale parameter λ.
9 The beta negative binomial distribution of the degree k, where the law of distribution ν is approximated by the gamma distribution with the form parameter m and the scale parameter λ.
10 The modified Poisson distribution, where ν is distributed by the normal law and estimations of parameters m_ν, S_ν^2 are known (volume of sample $k < 10$).
11 The extreme distribution, where ν is distributed by the geometrical law and the mean estimation ν is known.

The given one-dimensional parametrical models of distribution do not solve problems of estimation and forecasting of non-success risk or accidents. We can do uncountable quantity of curves through points and it is difficult to say which curve is better. Many works appear which are devoted to more sophisticated apparatus of one-dimensional analytical distributions and take into account "heavy tails" of distributions. The applications using "heavy tails" are also practically absent.

The main flaw of the classical theory of probabilities is by using virtually only one-dimensional distributions, that is, the influence of many factors (their real number reaches 100) is not taken into account. Let us consider this defect in more detail.

Multidimensional distribution in the probability theory. In real systems, the risk depends on many factors. For example, the security portfolio includes tens of valuable papers of different yields and risks. Often, different factors have different dimensions; the laws of distributions of factors are different and not normal.

Now, there is the mathematical theory only for multidimensional normal distributions, that is, each individual factor is distributed normally and its mean value and dispersion are known. The dispersion matrix of all factors is also known. The theory of calculation of the risk in real multidimensional systems, with influencing factors having different non-normal laws of distribution, is not created yet. This is one of the unsolved problems of mathematics.

The probability theory, as the applied science, is also named the "urn" theory, because the basic scientific results were obtained in experiments with urns and spheres of different colors. In those experiments, the probabilities or relative frequencies of random choice of different combinations of spheres of different colors were estimated. Thus, connection between the probability theory and the combinatorial theory was found.

Here, two things are important. First, the analytical formulas for the estimation of event probabilities appeared for convenient estimation, but basically were not obligatory, as almost all results could be obtained without these formulas, having the table of records of previous random samples.

Second, the combinatorial analysis has not obtained dule development because of astronomical number of combinations. The logic and probabilistic calculus can help combinatorics. We have got modern computers and algorithmic methods for solving difficult problems. Certainly, it is necessary to prove the basic results of combinatorics by the "large" computations. However, for applications, the astronomical computations are not obligatory if the appropriate numerical methods and algorithms for PC are developed.

Below, in different sections of the book, we shall show opportunities of the logic and probabilistic theory of modeling and analysis of non-success risk in complex systems. This theory confirms the rule "Razor of Occam" and concepts by John von Neumann and Norbert Wiener on applicability of the differential equations for the description of risk in complex systems.

6.4 The Human Being and Risks

The Science and Engineering are crystallization of wisdom, but often wisdom brings seeds of insanity and, therefore, rapid stormy development of a science and engineering results in a wide spreading of scientific and technical evil (Albrecht et al., 1995; Susumu Sato and Hiromitsu Kumamoto, 1995; Guding et al., 2001). Growing destruction of the environment everywhere over the world, accumulation of nuclear waste products, disasters, such as AIDS, failures and accidents in engineering, crises in economy, the political and information terrorism, etc., are symbolizing this fact. Modern industrial civilization or society of automation and information is characterized by fetishism of money and sciences, alienation of people and growth of dementia.

A human being and his participation as a risk element stand in the center of typical disasters in the modern world.

6.5 Frauds in Business

Let us consider some statistics on frauds in businesses in the USA (Albrecht et al., 1995; Solojentsev et al., 1999). The Chamber of Commerce informed that losses due to wastes of hired workers are estimated to be$20–40 billion annually. The volumes of such stealing go much ahead of those by house-breaking, hijacking, robberies and the usual thefts being counted in sum. Federal services estimate the common annual damage by swindle to be a sum from $60 up to $200 billion. The losses due to telephone swindle in 1991 in the market were estimated to be$10 billion.

The Federal Trading Commission (FTC) and the American Association health insurance (AAHI) estimate 10% as the number of fraudulent accounts on health services in the field of public health services. By the end of XX century, fraud in this area can cause more than $160 billion losses. By another estimation, fraud is absorbing up to $75 billion of all expenses in the USA on public health services.

Scientific researches show that three of ten workers are looking for possibilities to steal something, three others of ten will steal as soon as they have opportunity, and only four of ten will stay fair in any circumstances. Each year in the USA, 200 million thefts in shops (of goods in total sum $11.6 billion) occur.

According to Bureau of Internal Revenue of USA in 1990, the federal government got only 415 of all taxes. This underpay of taxes made $100 billion arrears. More than 660 ways of evasion from payment of taxes were disclosed.

In the USA, Russia and other countries, swindle has begun one of the main problems of economic safety of the state.

6.6 Errors of Personnel

It is notable that not only failures of technical elements of system result in accidents. A cause of interruption of normal operation may be a single unintended wrong action of an operator or a single omission of a needed action (the so-called errors of the personnel); it can also be a combination of technical failures and errors of the personnel.

In history of atomic power stations, more than 50% potentially dangerous incidents (small infringements in work) occurred due to mistakes of the personnel (*Risk Management*,2000). It should be taken into account that nuclear stations follow the program of quality maintenance and the required degree of quality of the equipment, the personnel and auxiliary devices of the "person-machine" dialogue is achieved. The concept of auxiliary devices of the "person-machine" dialogue includes the necessary and sufficient operator's devices for reliable and safe control of the power unit.

Failures of refuel systems of launching rocket systems can be classified as follows: 70% of failures were caused by aging and deterioration of the equipment as a result of long exploitation, 11% of failures were due to mistakes of the personnel, 5% occurred because of constructive defects, 11% were by exploitation factors and origin of other 3% is unknown (Krasnov, 2002).

Since the human being is often "a weak component", state of CS and its safety quite often cannot be estimated without taking into account quality of the personnel and working conditions in CS.

6.7 Asymmetric Actions of Terrorists

The sense of asymmetric actions of terrorists consists in making the greatest harm with the least expenses (amidst the expenses, terrorists count their own lives too). It is sad but today acts of terrorism on potentially dangerous objects and in places of mass accumulation of people are feasible.

Now and in the foreseeable future, as acts of terrorism in the USA (attacks on the World Shopping Center in New York and Pentagon in Washington), in Russia (explosions of buildings by the Chechen terrorists), in Israel (explosions in public places by Palestinian terrorists), etc., have shown, the mankind is vulnerable for the small radical groups, that are ready to play "not fair".

6.8 Hackers Attacks to Informational Networks

Now, we cannot imagine all gaps in the protection of our civilization. For example, the level of safety of global computer networks with the occurrence of computer viruses changed drastically. The racing of more perfect viruses against more effective anti-virus programs is going on. As the role of information infrastructure grows, the given class of risks can become more important. Dangers and risks can proceed from a person – hacker, not being stipulated by any technological necessity.

6.9 Personnel in Modern Civilization

Here, we present the results of the analysis of personnel work in the modern civilization, given by Sato Susumu and Kumamoto Hiromitsu in their book "Re-engineering the environment" (Han and Shapiro, 1969).

Black boxes. Personnel are also becoming system components like robots for process un-automated on economic grounds. For example, the chemical plants are automated; all physical and chemical reaction processes are divided into unit operations or processes. Each unit operation is considered as a black box, automated and all unit operations are then integrated and controlled. Of interest are input and output relations for each unit operation, and the internal mechanisms of the operations are often neglected. The control over a unit operation is performed on the basis of various measurement variables such as temperature, pressure, rates of heat generation and stream flow rates. The unit operation looks like a black box to the human operators. Automation has increased the number of black box systems. This inevitably increases the risk of accidents, due to incomplete understanding of processes inside the black boxes or of the ways of interaction between the black boxes.

Automation is fully based on modern rationalism which

- subdivides the whole into elements;
- neglects qualitative aspects of objects;
- recognizes objects by quantities.

Each element thus becomes a target for investigation, the elements are integrated to form the whole, and the resulting is controlled by computer. Real objects, however, have qualitative and quantitative aspects, and the automation cannot fully represent real processes.

Human errors. Automated manufactures require control and monitoring from the control center as well as daily inspection and maintenance of each elementary process. The automated systems are designed in such a way so as to monitor each process by control panels in the control room. However, the machine and process may sometimes cause abnormal events which cannot be monitored from the control center. When these events are overlooked, serious accidents may occur.

Consider a chemical plant where unit processes are connected by pipes. Assume that high-temperature and high-pressure fluids (or gasses) flow through the piping network. Such a chemical plant has a high risk of small leakage of fluids. The leaked fluids may accumulate, and a spark can cause explosions. This type of leakage cannot be detected by indicators on the control panel and daily inspections are required.

Operator errors are inevitable for current automated systems. Monitoring tasks are monotonous, boring and leading to loss of concentration. Humans are not good at this type of monotonous work. They find more satisfaction in tasks which require judgments to adapt themselves to changing environments because such tasks lead to learning by experience. Monitoring tasks with such lack of stimulation are confidence-destroying and error-prone. The human errors do occur frequently in modern automated systems. And the errors symbolize unacceptance of the monotonous monitoring tasks.

A system or a subsystem is shut down by safety devices when a stable technological process is disturbed by operator errors. Failed components and other consequences of the accident must then be repaired to resume operation. Human errors are also characteristic for the processes of shutdown, repair and resumption. In automated manufactures, experience and expertise are minimized, types of labor are standardized and the number of expert workers is decreased. Thus, it is difficult to find people to cope with failures and malfunctions. Engineers have less experience in preventing unexpected chain-initiating event from developing into a large accident because each engineer is engaged in desk designs of small portions of the automation system. This fragmentation of knowledge may also be imposed by management, so that an engineer or scientist cannot go off on his own and start a competing design or construction company as he only knows a small part of the complete process.

Automation and intelligence. Some people suppose that automation increases the ratio of scientific or intelligent labor to manual labor. Others claim that blue-collar labor comes closer to white-collar labor by automation; blue-collar workers are replaced by gray-collar workers who are engaged in monitoring tasks; white-collar workers have risen to manage personnel and materials. It is said that automation requires intellectual labor which can only be performed by people with education levels higher than high school graduates.

An opposite view claims that gray-collar labor is literally gray because a stimulus challenging the labor disappeared. It is difficult to improve the human capabilities through gray-collar labor. The monitoring tasks make the nerves atrophy, causing a new form of fatigue unbearable for human beings. Modern labor-related medicine has pointed out that:

- optic nerves cannot sustain long periods of focusing on flat monitor surfaces;
- extensive periods of monitoring may yield diseases such as autonomic ataxia (loss of muscle coordination).

Therefore, monotonous labor typically observed in modern automated manufactures is no less inhuman than severe physical labor. The transition from blue- to gray-collar labor does not imply a transition toward more intelligent or more humane labor. The increase of workers with higher education has nothing to do with the ability or the level of intelligence of labor. The tendency of common higher education is a fashion induced by a longer lifespan, rather than a result of a conversion from heavy-type industries to a light-thin-short-small type of production.

It may seem that system programmers have the best work, as they are the brain and the center of automation of any manufacture. It is the case at the stages of development and implementation of new projects of automation. But after the project is finished, they are forced to leave the manufacture or remain for support of automation system and to perform boring routine work, and may be to kill time by writing viruses, or by another hacker's activity.

Management intensification. As meaningless, inhumane and isolated labor increases, management is being intensified. In the traditional steel production, management lines were not separated from the technological lines. These two types of lines were united into a technology/management system. Technological skills were important in these factories, and management was performed by various types of technological experts.

Clear separation of managerial and subordinate work is observed in recent reports on the steel industry. In Japan in a steel industry, a shift supervisor is a key person. He, as a "steel man" by definition, has to manage shift members not only at the factory but also at their homes. Monotonous monitoring tasks granted by only the nervous tension, subordinate tasks controlled by a timetable under the mask of scientific management and increasingly intensive labor drive the shift workers to despair. Worker's feelings are summarized by representative comments like "It turns out that I am now working three times harder than be-fore". Shift workers are being eroded by the labor intensification; their family life disintegrates, which, in turn, cause harmful influences on the worker's performance. Scientific management by the shift supervisor is no longer sufficient. He controls the life-styles of subordinates after working hours by making the excuse that he is taking care of their families. This style of management is required to push workers to work under conditions which make them lose their stimulus to work.

Increasing of routine workers. Automation in steel industries has created various types of routine labor while retaining some types of routine physical labor. The total number of workers has been decreased by automation. However, the number of routine workers increases considerably in subcontract factories. Automation results in the increase of the percentage of routine workers. Similar situations are observed in other industries. Rapid automation is in progress in car industries where the number of routine workers increases in assembly lines which are difficult to automate.

Some people predict that in the future, every process will be automated; they consider the current automation as a transition stage. It should be noted here that automation replaces routine tasks by machine operations only when such replacements are cost-effective. Some tasks are still difficult to automate. Besides, automation itself creates new types of routine tasks around human–machine interfaces. Computerization increases data input tasks at the input side, and data monitoring increases tasks at the output side. Automation results in the reduction of old types of routine tasks and growth of new types of such labor. It is notable that in automation, the total number of workers decreases, but the percentage of routine workers increases.

Third and fourth levels of industry. The reduction of the labor population in secondary industries (mining, construction, manufacturing, etc.) increases the number of labors in the tertiary industries. The development process follows the transition of accent from:

- primary industry (agriculture, forestry and fishing industry) to
- secondary (mining, construction, manufacturing), then to
- tertiary (commerce, distribution, transportation, communication, public relations, education, services), and finally to
- the fourth level (banking, insurance, real estate).

The expansion of third- and fourth-level industries is not a social need but a result of over-saturation of the second-level industry with labor population. The expansion of third- and fourth-level industries is evidenced by the flood of various types of advertisement, persistent and irrelevant enticements to buy goods, and excessive numbers of shops, banks and insurance companies. This inflation yields a transition of workers types from blue to gray and then to white collar. Some people claim that human labor has become more intellectual and less physical due to this transition.

Consider as a typical example a Japanese city bank which is a center of the money market. Today, the city bank is a leading company, but the labor in the bank is not challenging. Many white-collar workers are engaged in the counter services. The cashier at the counter continuously counts money received from customers. The money is handed on until eventually a final worker in a cash desk receives it. At some point of this process, the amount of the money received is printed on a bankbook, relevant data is sent to a host computer via communication link and the data is processed and stored in the computer. Money withdrawal follows a reverse process. Most bankers are thus doing routine jobs around the computer. Other bankers repeat routine home public relations (advertising) visits. The bank workers are seemed to be a bright group of white collars, but their jobs are unattractive and many bank workers have resigned from their companies.

The third- and fourth-level industries require many "key punchers". This job requires physical labor because it involves data entry via keyboards. The job uses mental labor because it reads a computer program list. However, such a physical or mental job is restricted to an extremely narrow domain. Such job of "key punchers" results in the inflammation of sheaths of tendon of wrist and autonomic ataxia and proves inhumanity of this job.

6.10 Principles of Risk Management in Design

Occurring incidents are: failures, non-successes, accidents and catastrophes. We appreciate the risk as the non-success probability, damage and the admitted risk. Principles of risk management, which shall be considered below, at the design stage are applicable with some success for technical, economical and organizational systems.

At the design stage, the system project in the form of the appropriate documentation is created; problems of risk management at stages of development and operational tests are solved, and the corresponding programs of tests and the monitoring system for the operation stage are developed.

At the design stage, the scenarios of danger of the whole system and its components are developed and analyzed, and structural, logical and probabilistic models of risk are constructed. The graph of dangerous states is built from the top – the final failure event or accident. Derivative events and initiating events are introduced. The possibility of localization of dangerous conditions at their occurrence is taken into account. At the given or chosen probabilities of initiating events, the risk of failures and accidents is estimated. It allows us, as a result of modeling and risk analysis, to choose constructive, technological and structural decisions for the achievement of acceptable risk.

6.11 Style, Concepts and Methods of Designer

Let us consider style, concepts and methods of the chief designer in providing safety and the minimal risk of a created new product. As an example, we shall describe the style, concepts and methods of work of the well-known aircraft designer A. N. Tupolev (Andrey Nikolaevich Tupolev, 1988), the founder of known airplanes: ANT, Tu-2, Tu-16, Tu-104, Tu-114, Tu-134, Tu-144, Tu-154, etc.

Style. The thinking method by A. N. Tupolev always corresponded to the level of the chief designer. He did not lose the common view, did not change into the narrow expert and also did not miss those details which are determining ones. He thought that the chief or main designer is appointed to realize the main idea. The head position obliges him to protect this idea from encroachments, from uncountable "corrections" under which pressure the idea can be simply buried.

The chief designer, who does not make great demands to people, cannot make a machine quickly and with a good quality. A. N. Tupolev strongly influenced the origin and development of the aviation science. First of all, it was revealed in his exclusive strictness to the authenticity of test materials, and also in irrepressible aspiration to understand the essence of considered processes and the physical sense of the investigated phenomena.

A. N. Tupolev accepted and realized only those new ideas that had strong scientific and technological basing. "I am not interesting, that you think. I am interesting, that you have learned, seen, understood, that you have made".

Concepts. First of all, the skills are to find and use among a set of the new ideas, providing progress, those which can be realized at this moment. It is well-known that the great Leonardo da Vinci put forward a big number of new and progressive ideas including the helicopter, parachute and ball-bearing; however for their realization, the mankind required five centuries.

In the project of a new experimental plane, Tupolev had to make a choice between two new engines: one of them was easier, more economic and less overall; another was worse on all these parameters. However, Tupolev chose the second engine. The reason was as follows: the first engine would demand reorganization of large number of oil refineries, and it would take time and require huge expenses.

The best bomber of World War II Tu-2 had a number of new features that it was possible to realize during war. The experimental plane by designer V. M. Illjasischev was constructed at the same time. It was a very good plane but it contained new solutions which could not be realized at the existing technological level of production; therefore, the plane remained experimental.

For some years, Tupolev searched a solution what a heavy jet bomber should be. Two interesting ideas allowed the project to be realized: A. A. Tupolev proposed to arrange engines behind a wing closer a fuselage, and A. A. Judin proposed to retract undercarriages in special gondolas on a wing.

Young experts were often amazed with his apparent inertness in questions of introduction of new, progressive proposals. The skilled people also understood that refusal of introduction of innovations on the final design stage helped to speed up development and to implement the new experimental object. There are examples when groundless use of innovations resulted in the creation of "crude" designs; their debugging strongly delayed introduction in mass production and in exploitation. A. N. Tupolev not simply refused innovations, but placed them in "port-folio" and used in the following development.

Any of his variant of a passenger airplane which borrowed elements of design of a previous military airplane comprehensively checked up in exploitation.

Methods. In the work of the chief designer, most of the time is taken by the organization of obtaining information on failures, by analysis and elimination of failures at all stages of life cycle of the plane.

Even the construction stage of the experimental plane and its units already brought the new information, including negative one: technological non-effectiveness, excess of the given weight, insufficient strength, defects and fairies of the equipment and mechanisms on test benches and input control.

He thought that only by full-scale tests and checks of parts and elements of the future airplanes on test benches, it is possible to find the confidence in reliability of the made decisions. Full flight airplane characteristics, characteristics of stability, controllability, maneuverability and fighting qualities of airplanes were determined at the stage of the state tests. The following order of classification of defects by results of the state tests was established. All defects were divided into four groups. The first group includes defects which are dangerous to flights and unconditionally needs immediate elimination. The second group is the defects which are harmless for flights but complicating the job and not allowing normally to perform tests. The third group is the defects, allowing to perform tests, but needed unconditional elimination on the tested plane. The fourth group is the defects, requiring elimination on prototype serial plane, with obligatory checking of the performed actions at the following flight tests of the prototype serial plane as a standard for series.

At exploitation phase the new information about the airplane as statistics of failures and destructions due to industrial or constructive defects and defects on unknown earlier reasons. The defect might be new kinds of resonant fluctuations in structures and systems, fatigue failures, unexpected increase of forces acting on levers of control or interaction in electric circuits and hydraulic systems and many others, including problems in the products of suppliers. Failures and accidents during mass exploitation, as a rule, are caused by a combination of defects of engineering (construction or manufacture) and operation (errors of personnel or flight crew). A. N. Tupolev made "uncompromising fight" for the correctness of reason analysis of accident and the defects, requiring elimination by the producer and the customer.

6.12 General Scientific Knowledge in the Area of Risk

General knowledge in risk area is the basis for designing and management of risk and safety of complex technical, financial and organizational systems (Vasiliev and Solojentsev, 1978; Solojentsev, 1991). The scientific paradigm proceeds from inevitability of failures and accidents and determines the necessity to use the concept of acceptable risk. To general scientific knowledge in the area, we shall refer the following principles:

- Errors in projects of CS are inevitable.
- If there are stocks of nuclear, chemical, biological and other energy, there are also the ways for their leakage to the environment and for the occurrence of accidents.
- If there are money, material and stocks, they can be stolen.
- There is no profit (in business) without risk.
- Everyone can swindle under pressure of circumstances, if valuables are badly guarded and it is possible to hide the trickery for some time.
- It is impossible to manage the risk without quantitative measurement and analysis of risk.
- Designers of system should think on its normal functioning, but they should also consider it from positions of the saboteurs looking for ways of its destruction (I. Rjabinin's rule).
- Chief designer of system should not apply all possible innovations at once; debugging the system in this case is practically impossible (A. Tupolev's rule).
- It is necessary to borrow reliable and well-tried elements and decisions from other products, companies and countries (G. Guderian's rule).
- It is necessary to minimize the variety of accepted decisions because it is impossible to provide high reliability for considerably heterogeneous systems (the unification rule by E. Solojentsev).

If we do not acknowledge these rules as scientific axioms, it is impossible to construct appropriate technologies for risk management. We note that many of these positions were rejected and not fixed in standards and laws of Russia. It was a common opinion that it is possible to ensure zero risk in any systems; besides, faultlessness and usefulness of activity of conscious people were supposed.

Let us make some comments on G. Guderian's rule. It appeared during World War II. When Soviet tank T-34 had proved its power in the actions, the special commission headed by G. Guderian made a conclusion that the German industry cannot create a similar tank for two years because for this time, it was impossible to create the fuel equipment for the diesel aluminum engine. The decisions and technology could only be borrowed from the enemy.

7 Models of Risks in Environmental Management

7.1 Models and Rules

The models, which are necessary for risk management at designing of complex systems, as a rule, are not described by differential equations. Mainly, the following types of models are used (Vasiliev and Solojentsev, 1978; Solojentsev, 1991; Solojentsev et al., 1999):

- scenario model of risk;
- structural or graph-models of risk;
- logical risk models;
- probabilistic risk models;
- models of trouble forecasting;
- models of testing technology, consisting of procedures and operations;
- models of objects description in the form of requirements of technical specifications;
- models of expenses of means for decisions and possible damage in the absence of decisions;
- models of organizational management in the form of frame networks, providing support of technology and protocols, release of reports and notifications;
- models of programs and test reports, represented by tables;
- models of states of system in exploitation in the form of the table "Conditions and parameters".

An information technology of the system designing should provide a convenient representation of these models and their communications through a database. Risk management at designing is also provided by expert systems with rules "if-then" and "by analogy", semantic and frame networks. For example, the technologies of designing of automated debugging and operational tests have procedures of forecasting, modeling, planning and decision making which contain about 100 elementary operations. From them, about 1/3rd operations are operations of documenting, about 1/3rd operations are operations of calculations on models and about 1/3rd operations require intellectual support for decision making with the use of rules.

7.2 Occam's Razor

In conditions of uncertainty or incomplete information for the description of complex natural and social phenomena, it is necessary to choose one of two or several theories (methods). For resolving such questions, scientists and experts should know the so-called rule "Occam's razor" named after the philosopher William Ockam (Guding et al., 2001). The meaning of this rule is that more simple explanations of some phenomenon are correct with high probability, than more complex hypotheses. In other words, if we have two hypotheses, explaining the same phenomena, it is necessary to choose that from them which includes the least number of assumptions and difficult calculations.

The sense of the metaphor, used to name the specified rule, is in cutting off superfluous principles and in constructing the model with minimal possible number of assumptions. "Occam's razor" is extremely useful but rarely used methodological tool. Nevertheless, it is necessary to notice that it is a philosophical principle, which is not true in all cases and, therefore, it should be applied with some care.

For example, consider the problem of approximation of ten points, solved to tare a spring. Should we use a straight line or any curve from practically infinite set? Application of Occam's razor results in the choice of the most beautiful and economical decision: the straight line is simple, then a complex curve.

Similar problems are usual in science and its applications, including modeling and analysis of risk of accidents and failures. The majority of models of accidents and failures, published now in scientific journals, are described by systems of differential equations, whose origin is not clear, or use the catastrophe theory or the chaos theory, or enter "abstruse" distributions of probabilities of casual events and "special" descriptions of "fat tails" of such distributions, etc. Above-mentioned approaches, along with the demonstration of mathematical methods, usually, do not give anything practically useful and it is impossible to check up these models.

At the same time, there is the simplest solution or "straight line" which consists in constructing a logical scenario of connection of events and writing down on the basis of the scenarios the logical and probabilistic risk functions of accident which, with the large simplicity and clarity, will show as all the processes can occur and at which values of probabilities. It is necessary to compare all other proposed models with "straight line", but often it is not done. Outstanding mathematician of the modernity von Neumann, one of the founders of cybernetics, discrete mathematics and the theory of computers, claimed that the mathematical methods, which would be developed for their application in social sciences, would be based on the logic, the combinatorial theory and the set theory rather than differential equations.

7.3 Physical Approach

The physical approach to debugging and operational tests, monitoring and diagnostics, risk management and safety of CS consists in the estimation of functional system abilities and their elements based on measurement of parameters and calculation of risk (Ivanchenko, 1974; Vasiliev and Solojentsev, 1978).

The measurement of parameters and indicators is applied to physical destruction (durability and wears), behavior of personnel, economic parameters, ecological parameters and the accident and failure risk. The physical approach provides a rapid estimation of functionalities of systems; the approach is alternative and complementary to the accelerated and durable tests resulting in failure or accident.

The physical approach provides an information communication of tasks of testing, monitoring, risk and safety of CS on the basis of the parameter measurement.

7.4 Scheme of Complex Object Management

Management of state and development of complex system, and also its testing and operation will be performed by us as complex object control of the control theory (Solojentsev, 1981, 1982); management consists in control of movement on the chosen program trajectory and correction at the deviation from it (Figure 7.1). As the parameter, specifying the trajectory, the risk can also be chosen.

Thus, the complex system is moved from the initial condition A to the given final condition B following the chosen program trajectory A-B divided into some stages $j = 1, 2, \ldots, \pi$. The correction is performed in case of deviation of system from the program trajectory.

Proceeding from this interpretation, the following basic concepts are introduced: $\Upsilon(\Upsilon_1, \Upsilon_2, \ldots)$ are controllable parameters; $H(H_1, H_2, \ldots)$ are stages; $U(U_1, U_2, \ldots)$ are controlling influences for the organization of stage; $Z(Z_1, Z_2, \ldots)$ are adjusting influences during stages. The named parameters are vector values.

Controlled parameters Υ are measured or observed parameters, which we use to judge about the system capacity to work. Leading at the first stage (point A) is chosen to be minimal in order to not destroy the system; the last stage (point B) is done on nominal or maximal functioning mode (mode with the maximal loading). The system is moved from initial condition to the final one through a finite number of discrete stages with progressive increasing of parameters.

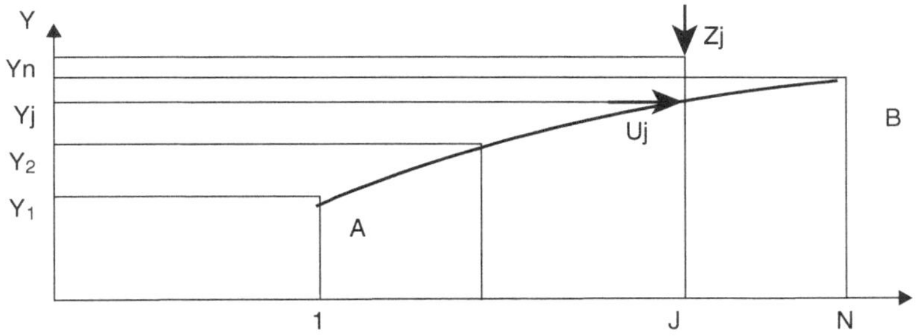

Figure 7.1 The scheme of control of complex object: Υ-controlled parameters, U-control actions, Z-corrective actions.

During development of the management program, designers beforehand prepare for possible accident by providing *Z-corrections*, which are certain variants of constructive or technological decisions or resources. In creating development program (debugging), it is necessary to determine values *Y*, *Z*, *U* for the stages of debugging *H*.

For complex system vectors *Y*, *Z*, *U* have big length and their realization can demand excessive means. For optimum choice of components of these vectors, it is necessary to know the expenses on:

- Measurements and management $Q_y\left(Q_{y1}, Q_{y2},\ldots\right)$;
- Control influences $Q_u\left(Q_{u1}, Q_{u2},\ldots\right)$;
- Adjusting influences $Q_z\left(Q_{z1}, Q_{z2},\ldots\right)$;
- Stages $Q_h\left(Q_{h1}, Q_{h2},\ldots\right)$;
- and also, the following possible damages if the actions are not made:
- $R_y\left(R_{y1}, R_{y2},\ldots\right)$ are damages in the absence of measurements and controls;
- $R_u\left(R_{u1}, R_{u2},\ldots\right)$ are damages in the absence of control influences;
- $R_z\left(R_{z1}, R_{z2},\ldots\right)$ are damages in the absence of adjusting influences;
- $R_h\left(R_{h1}, R_{h2},\ldots\right)$ are damages in the absence of stages.
- The scheme of management of a complex object is invariant concerning any objects. This scheme could even be used for the management of market transformations of economics in Russia on the program "500 days" by G.A. Yavlinsky.

7.5 Minimization of the Number of Decisions

Accepting as few as possible variants constructive decisions, it is possible to provide (at manufacturing and in exploitation) higher reliability of each accepted decision. We shall consider the designing problem with use as the criterion function "the minimal number of different decisions or the maximal unification of accepted decisions". The problem is of interest for a developing company, aspiring to satisfy needs of different consumers with more reliable products and smaller expenses, or for company, having high development and wishing to reduce too wide range of manufactured products and, accordingly, to reduce expenses for manufacture and increase the product reliability.

The formulation of such problem assumes the existence and possibility to find in the mathematical model of the product designing the set of admitted solutions (Vasiliev and Solojentsev, 1978; Solojentsev, 1991), exactly, the admitted discrete solution set (constituted, for example, by values belonging to series of sizes of the basic detail, established by the state or branch standard). The problem is solved in the dynamic optimization of standardized series of productions during performance of new projects. Beforehand, the time of an order receipt for products with new parameters and sizes is unknown.

Let us explain the problem statement using Figure 7.2. Here, the abscissa axis is the number (the basic size) of the product of standard series and the ordinate

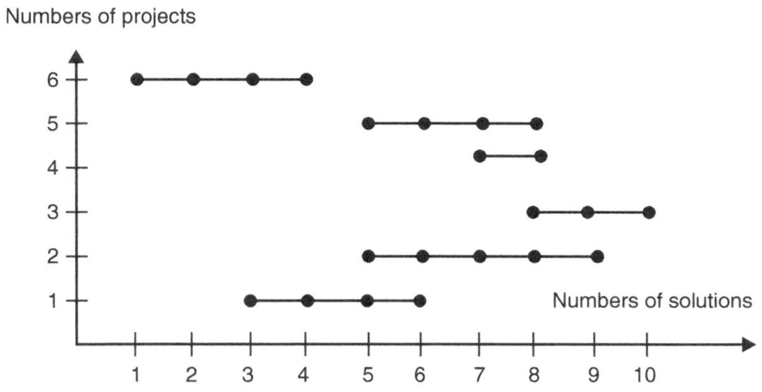

Figure 7.2 The scheme of dynamic training of dimension-type row.

axis is the number of the project in a sequence of its performance within, for example, two years under orders of the customer. The fat points on the horizontal lines are the admitted solutions for the project from the considered series. There are four allowable decisions for the first project, so the probability of the choice of the correct decision is equal to P_1=114. It is possible that the chief designer will choose decision 3 because it provides the minimal dimensions for the first project. Then making decision for the second project he should regret because the decision he has chosen and started in manufacture for the first project is bad from the viewpoint of unification. There are five admitted decisions for the second project; the probability of the choice of the correct decision is again small P_2=115, etc. Thus, depending on "luck" of the designer, for six projects, he could obtain from six different constructive decisions

$$\begin{aligned}&(1 \to 3, 2 \to 5, 3 \to 9, 4 \to 7, 5 \to 6, 6 \to 2) \text{ up to two} \\ &(2,3,4,5\ ) \to 8; (1,6) \to 3.\end{aligned} \tag{7.1}$$

Let us state the algorithm for solving this problem of dynamic optimization of series with the criterion "of small possible number of different standard sizes". Elements of established series of standard sizes $D_1, D_2, \ldots, D_n$ are given by normalized weights $C_1, C_2, \ldots, C_i, \ldots, C_n$ $(\Sigma Ci = 1)$. If the company only begins development and manufacture of projects, these weights should be fixed $C_1, C_2, \ldots, C_i, \ldots, C_n$ $1/n$. If the company has already developed projects, the normalized weights can be chosen, for example, proportional to the quantity of product release with the given standard size. As a criterion function, we shall use the following expression for the entropy of series:

$$I = -\sum Ci . \ln Ci_1 \quad i = 1, 2 \ldots, n \tag{7.2}$$

the meaning of the latter can be explained on the example of information entropy. Suppose that information either D_1, or D_2, or...or D_n, is transferred with the probability of these events C_1, C_2,...,C_n. If probabilities C_1, C_2, ..., C_n are approximately equal, it is impossible to give preference to any of events. In this case, one speaks that information entropy is small. If probabilities C_1, C_2, ..., C_n appreciably, it is presumed that the information having the greatest *a'priori* probability was transferred.

The entropy of series (7.2), as well as the entropy of information or thermodynamic system, serves for an estimation of a measure of "disorder" of series members. Maximization (increasing) of entropy of series corresponds to the increasing distinctions in weights between members of series; it means increasing manufacture of members of series with the greatest weights.

Thus, during designing of new products, it is necessary to choose from the admitted decisions one that, as much as possible, increases the entropy of series or, that the same, to choose as the decision, the element of series, which has the maximal weight. Naturally, the weight of this series member needs to be increased proportionally, and it is also necessary to make normalization of weights of all members of series and, by that, to obtain *posteriori* probabilities (weights).

The stated approach was used for designing of cylinders (characteristic size is the diameter) of piston compressors in the industry. The series of diameters (according to the standard for diameters of piston rings) consists of 108 members; the admissible decisions for the project were selected by solving the problems of linear programming, first on min, and then on max (Vasiliev and Solojentsev, 1978; Solojentsev, 1991).

The group of the projects having a place actually for last three years was presented to the weighed series of cylinder diameters on serial productions. The group of compressors of one of the companies could have, basically, 32 cylinders with different diameters. During consecutive development of projects, the designer chose cylinders with 24 different diameters; 13 different diameters were chosen by the considered method.

7.6 Structural Design

Structural designing serves the important purpose of designing systematization and increasing of object reliability. Structural designing is the method of modular designing of objects at which process of designing is represented as hierarchy of levels of comprehension of the object (Solojentsev, 1991; Solnitsev, 1991). Thus, each certain level is completely isolated from details of the lower levels (Figure 7.3).

The method of structural designing assumes that at the first stage of design, the project S is expressed in terms of its essential concepts. At this level of abstraction, we fix some objects (elements) of the second level S_1, S_2....These components are considered further as the object components, which will be decomposed at the following level. The process of definition proceeds up to the level where elements of the object become elementary and indivisible. Moving

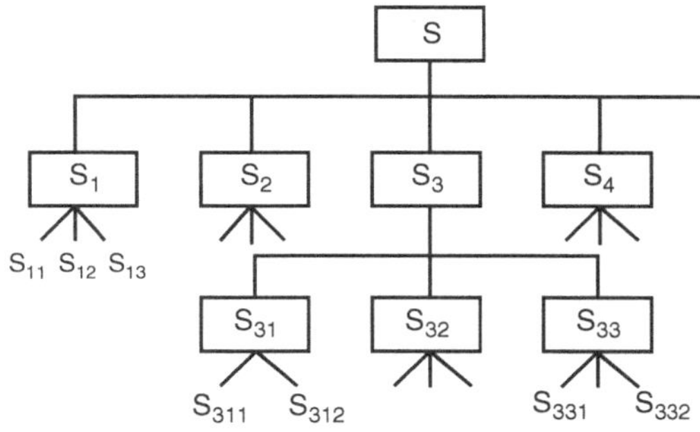

Figure 7.3 Scheme of structured design.

deeper into essence of the problem and dealing with the most complex object-elements, the designer abstracts details of the lower level. Therefore, it is possible to provide for modification of the object. The objects of lower functional level can have some alternative realizations. It leads to the problem of optimal choice or unification of decisions with the use of exiting wares of the world market.

Such sequence of the designing, named "from top to down", is one of the basic ideas of the structural designing. The second important idea of the structural designing is the use of simple and evident schemes of management of the project, which becomes foreseeable and "controllable".

The object development process begins from comprehension of requirements of consumer and market on the creation of original elements or borrowed elements from former designs, or from realized elements by specialized firms.

Certainly, during "from top to down" designing, the returns to higher levels are possible, if there are no effective decisions for elements of the lower level. However, it is not a reason to reject the basic ideas of the structural designing method.

7.7 Concept of the Acceptable Risk

The central philosophical question in the safety problem is the choice between the concept of "absolute" safety and the concept of "acceptable" risk. For the first time, the concept of the acceptable risk was stated in I. A. Rjabinin's papers. We shall describe this idea following one of his monographs (Ryabinin, 2000).

Substantiation of the concept. At first, more humanistic concept (at first sight) of the absolute safety was accepted. It was a basis for definition of the appropriate standards in the nuclear power and in other branches of industry. The detriment of the zero-risk concept is by the presumption that it is possible to exclude any danger for population and environment if we do not spare efforts and means for the creation of safety engineering systems and the serious organizational acts, providing a high level of discipline.

However, even the use of most effective safety systems, and advanced methods of monitoring in technological processes does not provide, and cannot provide, in principle, the absolute reliability of a system operation, excluding any accidents. The zero probability of catastrophes is reached only in systems with no reserved energy, chemically and biologically active components. On other objects, the catastrophes are possible; they are not excluded even with the most expensive engineering acts. The concept of the absolute safety is contrary to the internal laws of nature which have a probabilistic character.

The concept of the acceptable risk has many opponents. They consider it as immoral, saying that this concept gives designers the right to plan failures with probability less than the acceptable one. However, it is more immoral to mislead ourselves with hopes on the unattainable absolute safety.

Probabilistic risk analysis having been used outside Russia for many years has allowed to accept a set of new actions to increase safety of operation of nuclear stations and other potentially dangerous manufactures. The concept of the acceptable risk allows us more reasonably (with open eyes) to concentrate and distribute means not only for accident prevention, but also for preliminary preparing for emergency actions in extreme conditions.

Having agreed with the acceptable risk concept and necessity of probability calculation of danger of technical systems, it is necessary to choose suitable mathematical tools. Such tools, as a rule, are the probability theory, mathematical statistics and mathematical logic. The development of the logical and probabilistic safety theory (LP) of CS is seemed to be very perspective.

As the safety LP-theory, we understand the basic knowledge on calculations of the risk of failures and accidents in the structural complex systems. It is based on the logical presentation of development of dangerous conditions and mathematical methods for the calculation of the validity of functions of the logic algebra. The LP-methods of safety research allow us to reveal objectively the most dangerous places, the reasons and initiating conditions; the methods form another ideology of developers and induce experts to concentrate their efforts to the decision of principal problems.

Economic choice of the acceptable risk. The most general and universal method of calculation of the risk value (Sokolov, 2001) is the approach based on the economic analysis of safety. According to this method, the criterion of optimum of safety level is the minimum value Q that is the sum of two components: Q_1 (r) is the given charges for safety security with the risk r, and Q_z (r) is the direct damage, caused by the risk r. Thus,

$$r_{\text{opt}} = \arg\min Q(r) = \arg\min\left[Q_1(r) + Q_2(r)\right]. \tag{7.3}$$

The value raptness can be accepted as the acceptable risk value. The graphic illustration of the above-mentioned expression is shown in Figure 7.4.

The acceptable risk value depends on the national economics level. The higher is the economics level, production relations and safety culture, the higher is the level of requirements, made by society, to the safety of potentially dangerous

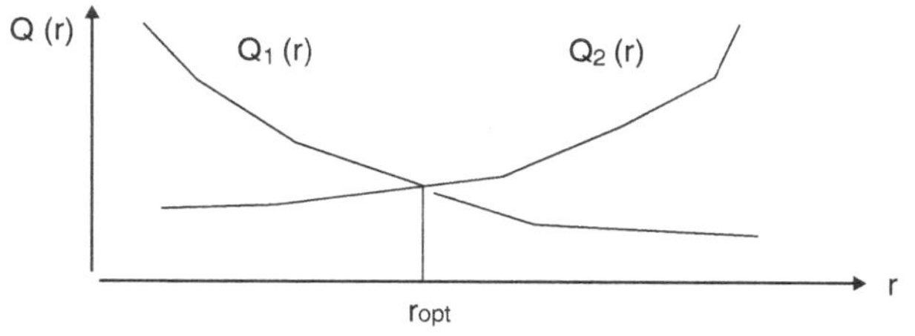

Figure 7.4 Finding of best value of risk.

objects, i.e. to the lower value of the acceptable risk. In the process of economics development, the requirement to safety should increase, and the value of the acceptable risk should reduce. Introduction of the failure risk as the universal characteristic of the safety meant, in some sense, revolution in theory of safety management.

7.8 Markowitz's and VaR-Approach to Investment Risk

Investments are the basis of the market economy in developed countries. The security portfolio theory is the most widespread modern theory of investments. It makes it possible to optimize, simulate, analyze the risk and operate by the security portfolio risk. It solves the problems of forecasting and optimization of yield and risk.

In Markowitz's theory and VaR-approach (Value-at-Risk),"models of averages and dispersions" are used (Markowitz, 1952; Sharp, 2001). For each security in a portfolio, the yield, as the mean of distribution, and the risk, as the mean square deviation and measure of uncertainty of yield, are taken into account. Concepts such as diversification, curves of indifference of the investor, available and efficient sets of portfolios are used. The normal distribution laws of yield both for each security and for total portfolio are used.

Problem of selection of an investment portfolio. The investor has a fixed sum of money for investment and wants to invest this money for a certain time interval. The beginning of the period is designated t=0 and t=1 corresponds the period end. At the period end, the investor sells securities which were bought. As the portfolio has some securities, making this decision is equivalent to selection of the optimal portfolio from a set of possible portfolios.

Making a decision in the moment t=0, the investor should take into account that yields of securities and the portfolio in the forthcoming period are unknown. However, the investor can estimate expected (or average)yields of various securities, being based on some assumptions and then invest money in securities with the greatest expected yields. Markowitz notes that it will be, in general, an unreasonable decision. The typical investor wishes the highest yield but simultaneously wants the yield to be so determined, as far as possible. It means that the investor, aspiring simultaneously to maximize expected yields and minimize uncertainty (the risk), has two purposes, contradicting each other, which should be balanced at purchase decision making about at the moment t=0. Markowitz's approach for decision making makes it possible to take into account both these purposes adequately. The approach implies diversification, i.e. purchasing of not one, but several securities.

The yield of security j for one period can be calculated using the formula

$$r_j = (r_{1j} - r_{0j})/r_{0j1} \tag{7.4}$$

where R_{oj} is the security yield at the moment t=0; R_{lj} is the security yield at the moment t=1.

As the portfolio is a set of various securities, its yield can be calculated analogously:

$$r_p = (r_1 - r_0)/r_{01} \tag{7.5}$$

where R_o is the portfolio yield at the moment t=0; R_1 is the portfolio yield at the moment t=1.

At the moment t=0,an investor cannot know the yield for all portfolios. Hence, the investor should consider the yield, connected to any of these portfolios, as a random variable. Such variables have characteristics, one of them is the expected (or average) value $\overline{r}_p$, and another is the standard deviation σ_p.

The investor should base the decision of the portfolio selection exclusively on the expected yield and the standard deviation. It means that the investor should estimate the expected yield and the standard deviation for each portfolio and then choose "the best portfolio", basing on the ratio of these two parameters.

Expected yield. As it was marked earlier, the portfolio represents some set of various securities. Thus, the expected yield and the standard deviation of portfolio should depend on expected yield and standard deviation of each security, included in the portfolio. Besides, obviously it should be taken into account what part of money is invested in each security. The expected yield of the security portfolio is as follows:

$$\overline{r}_p = \sum_{i=1}^{N} X_i \overline{r}_i, \tag{7.6}$$

where:

X_i is the part of money, invested in the security i;
$\overline{r}_I$ is the expected yield of the security i;
N is the number of securities.

Standard deviation. The measure of the risk should estimate deviations of the achieved result from the expected one. The standard deviation is the measure, allowing us to do it, as it is an estimation of the real yield deviation from the expected one. In the case when the yield distribution of a portfolio can be approximated by a curve of normal distribution, the standard deviation is really a very good measure of uncertainty degree for the estimation of portfolio trend. The approximation is often considered as the plausible assumption at the yield analysis of diversified portfolios when the investigated period of holdings of securities is short (for example, quarter or less).

The formula for the calculation of the standard deviation of a portfolio is as follows:

$$\sigma_P = \left[\sum_{i=1}^{N}\sum_{j=1}^{N} X_i X_j \sigma_{ij}\right]^{1/2} \tag{7.7}$$

where σ_{ij} is a covariance of security yields i and j.

Analysis of portfolio. In Markowitz's approach to the decision of the problem, an investor should estimate alternative portfolios from the viewpoint of their expected yields and standard deviations, using Indifference Curves. In the case when the purpose is to avoid the risk, the portfolio, laying on the Indifference Curve, which is located higher and more to the left than other curves, will be chosen for investment.

From a set N, securities are possible to combine the infinite number of portfolios. Fortunately, the investor should only consider a subset of the set of all possible portfolios, belonging to the so-called efficient set. The investor will choose the optimal portfolio from the portfolio set, where every portfolio provides the maximal expected yield for some risk level or provides the minimal risk for some value of the expected yield. The set of portfolios, satisfying these two conditions, is the efficient set or the efficient border.

The VaR-approach for the selection of portfolio (Value-at-Risk). The VaR-approach for the selection of the portfolio by a criterion of allowable losses (draw down criteria) is alternative to Markowitz's approach. We shall consider the typical case of an investor who is willing to avoid risk. The choice of the optimal portfolio is made by the condition of maximization of the admitted yield

$$R_{ad} = m_p - h_a^{*}\sigma \rightarrow \max; \tag{7.8}$$

taking into account that $\mathrm{VaR} = h_a^{*}\sigma$, and the latter formula can be written as follows:

$$R_{ad} = m_p - \mathrm{VaR} \rightarrow \max; \tag{7.9}$$

Here, m_p, R_{ad} are the expected and minimal admitted yields of security portfolio, respectively, ha is the number of standard deviations in quantile of order a (level of trust); for example, for trust level a=95%, the value of h_a is 1.65; and for a=99%, the value of h_a, is 2.33; σ is the standard deviation of the portfolio yield; VaR is a probable loss.

In conclusion, we note that the assumption about the normal law of distribution of every security and portfolio yield (according to Markowitz's theory and VaR-approach) is very strong and not always justified.

As it will be shown in this chapter, and last chapters the non-success risk LP-theory with GIE successfully solves the choice problem of the optimal security portfolio unifying Markowitz's and VaR techniques. It allows us to remove the essential assumption in the portfolio theory about the normal distribution law of the yield of each security and the whole portfolio and to solve new problems for the analysis and forecasting of the portfolio risk.

7.9 Active and Passive Management of Risk

Let us formulate the concept of safe deterioration of the material resource part (Pechenin, 2000). If a material resource in the start of usage has a value, exceeding the necessary resource for the operation of the object, then the process of resource deterioration has two stages. At the first stage, the remained not-depreciated part of resource completely provides trouble-free operation of the object. The probability of a failure, caused by the deterioration of the given resource, does not differ from zero. At the second stage, the material resource is depreciated much, so that the probability of failure of the object because of deterioration of the given resource accepts some non-zero. The time corresponding to the moment of transition from the first stage to the second stage is the so-called threshold time. It is possible to control the threshold time: the influence of any material resource on the non-failure operation of an object can be investigated.

The material resource can be considered as a set of resources, where each resource is wearing out in the course of time. For each resource, there is some function $\mathrm{Rs}(t, x_1,\ldots,x_m)$ that represents dependence on time *(t)* and on conditions of operation $(x_1,\ldots,x_m)$. This function is usually investigated beforehand by an experiment with material resource $_i$. In the range $t_0 \div t_1$, the function is not defined (Figures 7.5 and 7.6) which corresponds to the fact that frequency of failures as a result of deterioration of the given resource in the range $t_0 \div t_1$ is equal to zero. Since generally the material resource consists of a set of internal resources, there is always a danger of existence of the deterioration function of an unexplored internal resource $\mathrm{Rs}(t, x_1,\ldots,x_m)$ which has some non-zero finite values in the range $t_0 \div t_1$. A single fact of failure because of deterioration of an internal resource $\mathrm{Rs}(t, x_1,\ldots,x_m)$ leads to the necessity of researches which will change exploitation conditions in such a way that the repeated occurrence of

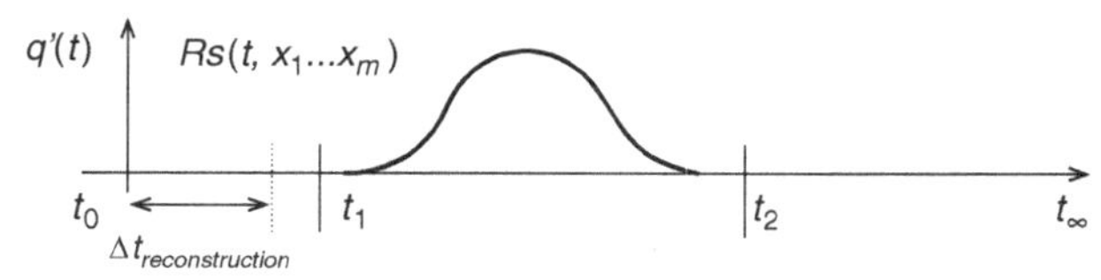

Figure 7.5 Frequency distribution of failures in result of wearing of material resource.

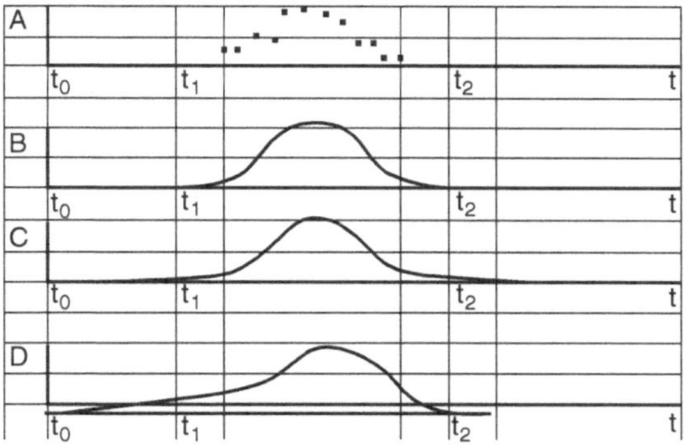

Figure 7.6 Different interpretations of frequency distribution of refusals as a result of wearing of material resource.

the given event in the range $t_0 \div t_1$ becomes impossible. Since to the left of the point t_l, the probability of breakdown is equal to an infinitesimal value, and to the right of the point, the probability has finite values, it is reasonable to suppose that the threshold time is near the point t_l. The traditional approach to the selection of the distribution function of failure does not assume existence of the threshold time.

Experimentally obtained data on the object breakdown, which happen because of deterioration of any controllable material resource, represent the operating time, laying in a limited range of time. The life of a material resource can be divided into three qualitatively different periods:

$t_0 \div t_1$ is the period, when the material resource is sufficient and there is some reserve of the resource;
$t_1 \div t_2$ is the critical period, when the material resource has no reserve and continues to wear out;
$t_2 \div t_\infty$ is the period, when the material resource is already exhausted.

The moment t_1 is the threshold time. If work with the material resource is correctly organized, that is beforehand, before the threshold time, the material resource is regularly restored, it is possible to control the threshold time. The value of the reserve factor for the time between two procedures of restoration of the material resource following one after another can be found on the basis of the investigated material resource and the value of a possible damage in the case of failure, caused by deterioration of the given resource. Hence, the opportunity to control the threshold time directly depends on researches, that is, legitimacy of concept of the safe deterioration of the material resources depends on the realization of timely and sufficiently wide program of researches of material resources.

In the case when the material resource is widely used, information on the resource deterioration is usually sufficient and available. Researches should be directed to revealing qualitatively new features of the resource. If the material resource is used seldom or is unique, researches can be productive only when the intensity of the material resource deterioration which can be achieved in researches is higher than the one in the real operation process.

Practical use of the concept of safe deterioration of the material resource consists in the following: before the threshold time, it is necessary to manage by the risk actively, raising reliability of the weakest and most dangerous elements of the system. It is necessary to use results of monitoring of conditions of these elements. After achieving the threshold time, it is necessary to replace dangerous elements or, if it is possible and reasonable, to ensure the elements or the system as the whole.

7.10 Algorithmic Calculations

It is not a secret that the most mathematics and mechanics are still demonstrating us your skill by analytical solutions and transforms and are using a computer for printing results of their intellectually refined excises. However, at quite "strong" assumptions, their problems come to analytical calculations for points, strings, plates, cases, etc.

Problems of logic and probabilistic estimations and analysis of risk are always connected with complex algorithmic calculations, which are so labor-intensive, that arise the problem of estimation of the complexity algorithm and decrease laborious calculations.

Algorithmic calculations are the following stages of constructing and using the LP-models:

- constructing of the risk L-function and the risk P-function;
- solution of the optimization problems with the calculation of criteria and characteristics of risk models, of risk elements, risk objects and the system in the whole;
- fulfillment of combinatoric and logical and probabilistic analysis of risk.

7.11 Arithmetical and Logical Addition

Initiating factors (signs) influencing the final event of the system can be added arithmetically or logically. The number of such added factors can be from several units to several tens. Below, we shall study the dependence of the final event probability from values of probabilities of factors and their numbers, and also, we shall compare results of arithmetical and logical addition of probabilities of sign-events.

The logical function for addition of events (factors) $Z_1, Z_2, \ldots, Z_n$ is as follows:

$$Y = Z_1 \vee Z_2 \vee \ldots \vee Z_j \vee \ldots \vee Z_n. \tag{7.10}$$

In words, it means that the failure occurs, if occurs any one, any two...

or all initiating events. After orthogonalization of the logical function (3.9), the following probabilistic function can be written (probabilistic polynomial) (Solojentsev et al., 1999):

$$P = P_1 + P_2, Q_1 + P_3 Q_1 Q_2 + \cdots \tag{7.11}$$

where

$P_1, P_2, \ldots, P_j, \ldots, P_n$ are probabilities of events $Z_1, Z_2, \ldots, Z_n$;

$$Q_1 = 1 - P_1, Q_2 = 1 - P_2 \ldots \tag{7.12}$$

The arithmetical function for the addition of events is as follows:

$$P = P_1 + P_2 + P_3 + \cdots + P_j + \cdots + P_{n1} \tag{7.13}$$

where $P_1, P_2, \ldots, P_n$ are weights of factors $Z_1, Z_2, \ldots, Z_n$.

The value P of the probabilistic polynomial (3.10) always belongs to the interval $[0,1]$ at any values of probabilities of initiating events $0 \le P_j \le 1$; $j = 1, 2, \ldots, n$.

If there is one sign-event *(n=I)*, the probability of the final event P in logical addition will linearly depend on the probability of this sign-event P_l (Figure 7.7). If there are two initiating sign-events *(n=2)* in the logical addition (3.10), the probability of the final event P will have *S-type* dependence on probabilities of sign-events P_j, j=1,2 (which are given with identical values). If there are three and more sign-events, the probability of the final event P will also have S-type dependence on probabilities of sign-events P_j, j=1, 2,... (which are given with identical values too). The steepness of S-curve will increase with increasing n.

The probability of the final event in the logical addition depends both on the number of sign-events and on their probabilities. The saturation of probabilities (P=1) also depends on these factors. We note that only low probabilities of initiating events (sign-events) provide the small total risk $\left(P = 0.02 \div 0.04 \text{ for } P_i = 0.001\right)$. Comparison results of logical and arithmetical addition of probabilities of sign-events are shown in Figure 7.7 for the number of sign-events 41, 20, 5, 1. For the

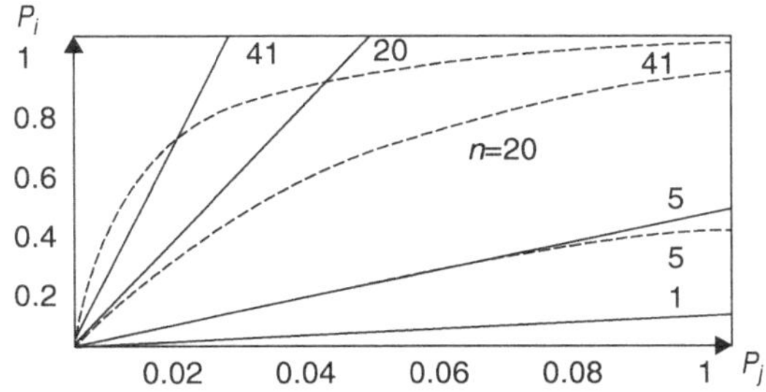

Figure 7.7 Risk in function of number and probabilities of initiating even.

big values of weights of signs $P_j, j = 1, 2, \ldots, n$, and for the big number n, the final event probability, calculated as the arithmetical sum of probabilities, becomes absurdly large $(P \succ 1)$. The arithmetical and logical sums are close only for small values of probabilities of initiating events and the small number of the events. Therefore, the technique based on arithmetical addition has satisfactory accuracy only for small number of signs $n = 1 \div 3$ and their small weights $P_j = 0.001 \div 0.0001, j = 1, 2, \ldots, n$.

Comparison of polynomials for arithmetical and logical addition shows that the logical and probabilistic polynomial has a more complex structure and, consequently, better possibilities for the adequate description of the final event risk.

It is also notable that the polynomial can replace with success a neural network (NN) with arithmetical addition of edge weights. The logical function for the probabilistic polynomial also has quite clear sense in comparison with NN. Formulas on the basis of NN are deprived of the physical and logical sense.

7.12 Risk Management in Operation on Basis of Monitoring

At the operation stage, an estimation of CS accident risk is made on the basis of scenarios by using results of monitoring. The monitoring allows us to estimate values of element deterioration, real loads and vibrations, operation features, readiness and conditions of the safe operation. The quantitative estimation of accident risk allows us:

- to analyze the CS risk;
- to accept the reasoned decision on prolongation of safe operation and service life of CS;
- to develop proposals on maintenance of safe operation;
- to organize the process of personal training for safe operation;
- to plan acting in dangerous situations.

Monitoring is the essential part of safety system management of complex technical, economic and organizational systems.

7.13 Destruction, Wearing and Deterioration of Equipment in Operation

For constructions and buildings, being in operation for a long time, the cause of accidents can be degradation of material properties, beyond-limit levels of stored damages, appearance of uncontrollable development of cracks, cavitation wearing, etc.

During a long-time operation, CS elements worn out and age. We observe corrosion, active degradation of equipment components, effects of corrosion-active substances, effects of dangerous and harmful factors such as higher and lower temperatures, humidity, vibrations of different spectrum and amplitude, etc.

Table 7.1 Characteristics of using design resource of objects

Level of resource usage	*Relative number of objects*
More 0.50	60%
Ur, to 0.75	20%
Up to 1.0	15%
More 1.0	5%

Combined simultaneous actions of the named factors result in the accumulation of rust, appearance of cracks in elements of construction and in welds, breakdown of air-tightness of flange packing, reduction of insulation resistance of cable lines because of aging polymeric covers, etc.

Potentially dangerous objects and manufactures, as a rule, have considerable usage of resource (Table 7.1). In the most crucial branches (power, petrochemical industry, gas chemical industry), the potentially dangerous objects have used a resource at a level of 75–90% (*Risk Management*, 2000). In the first years of XXI century, the amount of potentially dangerous objects with the indicated levels of used resource will increase by approximately 10%. Thus, the required expenses for the liquidation of consequences of extraordinary situations, modernization, renovation and withdrawal of the equipment will result in the reduction of gross national product as much as to 5–10%.

Examples. Today, many atomic power stations have high, above 65%, levels of wearing of basic production equipment. Insufficient attention is paid to modernizing, repair and preventive maintenance of equipment. By social causes, the industrial and technological discipline drops. In the chemical complex, wearing of production equipment comes to more than 80%, and about half of long-distance pipes have been in operation for more than 20 years. Repairing and re-equipment of worn-out equipment are much behind of the needs.

7.14 Monitoring in Engineering

In engineering, a typical one is the conflict between reached values of operating time of complex technical systems and rigid requirements to quality of their operation during operation. First of all, it concerns the CS with high cost, unique ones, with a long time of building (nuclear reactors of power stations and submarines, launcher of rocket systems, power equipment, etc.). Therefore, it is necessary to create new operation-saving technologies (Prohorovich, 1999; Krasnov, 2000, 2002). The monitoring, as information technology, is intended for the evaluation of technical condition of the CS and its resource, for decision making on prolonging resource and maintaining of safe CS operation with the prolonged resource.

The essence of the new approach to the safe operation consists in a large-scale application in industry of monitoring of technical condition of the exploited CS in order to obtain timely information about their actual condition and to make decision about the order of their further exploitation. Thus, the safety of exploitation CS is determined both by its technical condition and by readiness of personnel to supply successful and safe operation.

Monitoring of CS exploitation is the process of systematic obtaining and initial processing of information about conditions of CS elements, affecting factors of environment and the operational processes realized in the CS. Further processing of information is used for numerical estimation of non-success risk of CS with prolonged service life.

The theoretic basis of CS risk estimation with the help of monitoring consists in the following. First, we write scenarios and the risk LP-models of accidents which allow us to compute the risk if probabilities of initiating events are known. The properties of CS elements change with time as they wear out, age and come to ruin. Monitoring fixes these changes. For prediction of change of element properties, equations of mechanics can be used.

At this stage, we also build models connecting properties of elements and probability of their failure (changing properties of elements results in changing probabilities of their failures). These models can be physical-statistical, statistical and expert ones. Examples of construction and usage of such models are given in Annikeichik et al. (2000), and Annikeichik and Sokolov (2001) for components of refueling systems of rocket launchers. Models are built for corrosion damages of pipe lines, wears of movable components, aging polymer and rubber products and errors of staff at localization of dangerous condition of the CS.

Thus, we can calculate accident risk for each CS condition by using monitoring results, and make decision on possibility of its exploitation, prolongation of resource, required repair and replacement of components or on impossibility of CS exploitation because of high inadmissible risk.

7.15 Monitoring of Infrastructure of Rocket Launcher

The world experience of the space activity testifies to the fact that problems of risk estimation and safe operation risk analysis of elements of the ground-based space infrastructure (GSI) are urgent. It is also necessary to create well-founded methods of achieving required risk level (Annikeichik et al., 2000; Krasnov, 2000, 2002). Basic GSI elements are refueling stations, technical facilities, launcher, measuring stations and other similar objects.

GSI is the complex system including a set of objects of higher danger. Thus, the majority of GSI elements are now exploited on the basis of decision on the prolongation of resource and service life. The features of exploitation of GSI elements at the stage of prolonged service life consist in the following. First, the GSI exploitation occurs in conditions of degradation of components of the equipment and it is difficult to meet the requirements on reliability and safety. Second, in a number of cases, the GSI exploitation occurs in conditions of poor technical readiness of tools of safety supply. All this results in the decrease of reliability and safety of GSI operation that confirms the urgency of developments, directed to the creation of methods of risk analysis, safety maintenance and risk management of GSI exploitation. The methods should take into account the actual technical condition of equipment and functional condition of personnel.

7.16 Scenarios of Accident Appearance

For providing safe GSI operation, the following problems are solved:

- selection of the risk parameter which allows us to estimate the safety level quantitatively;
- normalization of requirements to the safe GSI operation in the selected parameter;
- development of models for the calculation of the risk parameter value of safe GSI operation;
- development of technology for obtaining of initial data for the models;
- development of methods for analysis and management of safe GSI operation.

For quantitative risk estimation of safe GSI operation, a vector parameter can be proposed. It is the vector of probabilities of appearance of possible accidents:

$$P \prec L \succ = (p_1, p_2, \ldots, p_i, \ldots, p_L). \tag{7.14}$$

Application of such parameter enables to formulate the requirements to safe GSI operation with taking into account the level of possible damage by various accidents. The GSI application has as its final goal the condition S_k, obtaining some useful effect C_p (for example, profit). At the same time, during operation of GSI, there always exists the risk of accident appearance and damage of people, environment, equipment of the GSI and space rocket. The condition S_p corresponds to the accident event and it is characterized by the probability p, and the damage W_p. The graph of dangerous GSI conditions is shown in Figure 7.8.

Following basic principles of the concept "the admitted risk" (Ryabinin, 2000), it is possible to state that the risk of safe GSI operation can be considered as admitted one in the case when the positive effect from the GSI operation is more than the effect by accident, that is, the inequality holds:

$$(1 - p_n) \cdot C_n \succ p_n \cdot W_n. \tag{7.15}$$

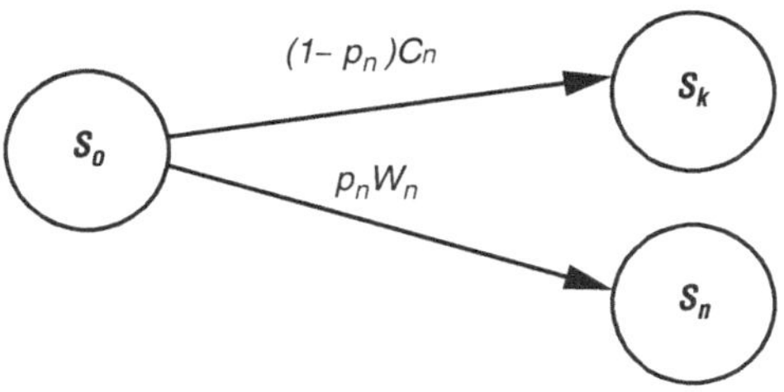

Figure 7.8 The graph of GSI state.

The inequality (7.15) reflects only the technic and economic aspect of maintenance of safe GSI operation and does not take into account the current condition of social relations or, speaking in other words, the level "admitted risk" currently accepted in society. For elimination of this flow, it seems appropriate to introduce the coefficient of admissible risk k_{ad}. Transformation of inequality with the coefficient k_{ad} allows us to obtain the maximum admitted probability of appearance of accident in the GSI starting from possible damage and expected useful effect from proper GSI application:

$$P_i^d = \frac{k_{ad} C_n}{k_{ad} C_n + W_{ni}}, \tag{7.16}$$

where W_{ni} is the possible damage by the appearance of 1-type of accident in GSI.

The full list of possible accidents on the GSI is found by the method of the morphological analysis. For example, for the space rocket launcher "Proton", the full list of possible accidents during preparation and launch of the rocket includes 66 items.

The quantitative estimation of the risk of safe GSI operation is made by the binary scheme (if the criterion is realized, then the demanded level of safety is ensured) and consists in the realization of criterion of suitability:

$$G : P \prec L \succ \in \left(P^d \prec L \succ\right), \tag{7.17}$$

where p^d<L> is the vector of acceptable values $\left(P_1^d, P_2^d, \ldots, P_L^d\right)$ of GSI accident appearance probabilities.

The criterion formally means:

$$G : \begin{Bmatrix} p_1 \le P_1^d, \\ p_L \le P_L^d, \end{Bmatrix} \tag{7.18}$$

The components of the parameter *P*<L> of safe GSI operation are probabilities $p_1, p_2, \ldots, p_n$ of appearance of accidents. They are calculated with the help of models of accident appearances in GSI. The basis of these models is the so-called "scenarios of accident appearances". They are prepared on the basis of models of the GSI exploitation process and formally described by logical and probabilistic functions of accident appearances.

The essence of the approach for the construction of models of accident appearances in GSI is the following. For the stage *i* of GSI operation, we determine all possible accidents, and for each of them, we construct the scenario of accident appearances. On the basis of this scenario, a logic function Υ_i of accident appearances is constructed. It permits, by using algorithms described in Ryabinin (2000), to determine the probabilistic function of accident appearances $P_l = P\{\Upsilon_l = 1\}$ expressed in terms of probabilities of initiating events (initiating events and conditions) of the scenario of accident appearances.

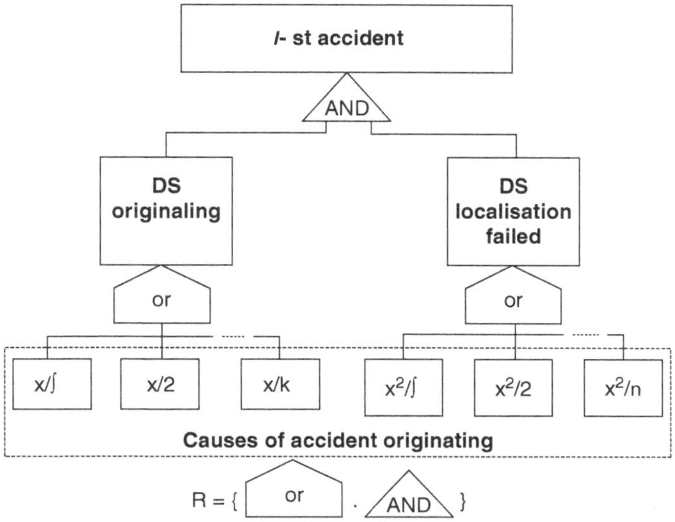

Figure 7.9 The scenario of incident originating.

The probabilities of initial events in scenarios of accident appearance are calculated as probabilities of their appearance at the considered stage of operation. In these models, the process of accident appearance is considered as consisting of two stages: appearance of the dangerous situation and its development as accident. Thus, we take into account the possibility of localization of the dangerous situation and prevention of damage. The example of the scenario of incident appearance is shown in Figure 7.9.

Main feature of the scenarios of accident appearances is the presence of internal "horizontal" connections. This dictates the necessity of application of algebra of logic for their formal description (Ryabinin, 2000); the causes of accident 1, possible at fulfillment of the considered operation stage, can be presented as events consisting in the failure of an equipment element (or an operator error) during operation of GSI at the considered operation stage. In the formalized form, these events can be written as follows:

$$\hat{x}_{lk} = \langle \hat{u}_{lk} < \hat{z}_{lk} \rangle, \tag{7.19}$$

where u_{lk} is the conditional (only within the considered operation stage operating) mean time between failures of the element k or mean time between errors of the operator; z_{lk} is the required mean time between failures of element k within the considered operation stage for its completion.

For determination of probabilities of the events, it is necessary to find distribution functions of random variables u_{lk} and z_{lk}, described with the help of physical-statistical, statistical and expert models (depending on the structure and quality of the data).

The application of the logic and probabilistic methods for the description of accident appearance scenario essentially simplifies estimation and analysis of risk of safe GSI operation. The scheme of such analysis with applying to one accident is described in detail in Solojentsev et al.(1999) and Ryabinin (2000); the meaning and features of the risk analysis of safe operation with applying to GSI and taking into account the selected parameter can be described as follows. First, a list of possible incidents of GSI is prepared. Further, the structural models of their appearance are designed, and the structural weights of each of the incident causes are determined (Solojentsev et al., 1999; Ryabinin, 2000). This allows us to form the ranged lists of causes for each accident. The latter show the parameters suitable for safety management. Besides, the LP-method allows us to determine the minimum cross-sections of the safe operation (or paths, which guarantee safety).

The possible accidents causes are divided into four groups: failures of GSI equipment elements, unrepaired errors of operators, non-success at localization of dangerous situations and initiating events. For each cause group, the standard models and recommendations on their application are proposed within the considered problem. For failures of equipment elements of GSI and non-success at localization of dangerous situations depending on structure and quality of initial data, one can use the physical-statistical, statistical and expert models. Thus, for physical-statistical models, the given data show the results of measuring main parameters; for the statistical models, the data denote the operation time until failure or the time of censoring elements; for expert models, the data show the expert interval estimations of corresponding parameters. For modeling of unrepaired eliminated errors of operators, the use of the known experimental models is proposed (Ryabinin, 2000; Solojentsev and Karassev, 2003); the initiating events of incidents are proposed to model with the help of the function with values 0 or 1, that is, the probability of appearance of initiating events can be equal to 0, or to 1, depending on the capability to manage them.

7.17 System of Monitoring

The basic tool for maintenance of the demanded safe GSI operation is the monitoring of parameters determining safe operation of systems during regular operation. Monitoring is understood as the process of regular obtaining and primary processing of information about parameters of technical condition of the system, which change influences safe operation of the system.

For the achievement of the purpose of monitoring within its realization, it is necessary to solve step by step the following problems:

1. to carry out the operation safety analysis of GSI means with the purpose of definition of parameters that require control;
2. to determine needed structure and volume of information about the checked parameters of elements;

3 to develop methods of control for the observed parameters and parameter monitoring strategy which include formation of the optimum schedule of estimation and analysis of safety by given criteria;
4 to conduct collecting of statistical and expert information on parameters determining safe operation of GSI systems and their checking by methods of nondestructive control;
5 to estimate and analyze the current safe GSI operation on the basis of obtained information. To forecast safe operation parameter values in the planed, pan of time with taking into account results of checking statistical and expert information on parameters determining safe operation of technical systems;
6 to elaborate actions on the maintenance of demanded safe operation of the considered system;
7 to update models of incident appearances in systems and the list of checked parameters and requirements of safe operation.

The system of parameters permitting to estimate safety of operation of technical systems is introduced. The two-level system of parameters is proposed for quantitative estimation and analysis of safe operation of technical systems. Vector of probabilities of accident appearances $Q<L>=\{Q_1, Q_2,\ldots, Q_L\}$ is introduced. Here, the sub-index designates the type of accident. The components of the vector $Q_1, Q_2,\ldots, Q_L$ are parameters of the first level. Appearance probabilities of incident causes (which, generally, can be element failures of a technical system, personnel errors, non-successes at localization of dangerous situation and initiating events) are denoted by

$$\left\{q_1^1,\ldots,q_\rho^1\right\},\left\{q_1^2,\ldots,q_\eta^2\right\},\ldots,\left\{q_1^L,\ldots,q_\zeta^1\right\} \tag{7.20}$$

and these are parameters of the second level. The probability values $Q_1, Q_2,\ldots, Q_L$ of incident appearances at technical system operations will be determined by parameter values of this level. We shall now consider in detail problems solved at monitoring.

The first problem is the analysis of safe operation of technical systems with the purpose of determination of parameters that require checking. In solving this problem, we investigate the operation process and the structure of technical systems with the help of the morphological analysis method and determine the full list of possible accidents at operation.

Then, the models of incident appearances are constructed on the basis of models of operating processes of technical systems. The basis of these models includes scenarios of incident appearances, described formally by logical and probabilistic functions of incident appearances. It is obvious that the probabilistic function of accident appearances can be expressed in terms of probabilities of the incidents causes, which are components of the vector

$$Q_\Sigma=\left\{q_1^1,\ldots,q_\rho^1\right\},\left\{q_1^2,\ldots,q_\eta^2\right\},\ldots,\left\{q_1^L,\ldots,q_\zeta^1\right\}. \tag{7.21}$$

After that, ranking of causes with regard to the risk factor is carried out, and the structural weights of the causes are determined. We assume that all causes have the same probabilities. Then, we find out the structural significance of accident causes possible at GSI operation. This is necessary to determine influence of the causes to the safety of operation of the technical system.

The second problem of monitoring is the determination of necessary amount of information on checked parameters, sufficient for their estimation, and development of instrumental base, necessary for obtaining this information. For this purpose, it is necessary to define the possibility to measure the value of the checked parameter or to define a list of indirect diagnostic characteristics permitting to determine its value. We use it for determining the structure of instrumental base needed for obtaining the demanded information.

The subsequent problem of monitoring of parameters, which determine safety, is the development of methods of checking the observed parameters. The problem solution requires implementation of the following steps:

- grouping elements by the type of checked parameters;
- definition of sets of checking methods permitting to control corresponding groups of elements;
- development of methods for checking of corresponding parameters for each group of controlled elements.

At realization of the first step, it is expedient to divide all set of controlled elements of technical systems into groups with controlled parameters of the same type. For each of the element groups, it is possible to compose a list of methods of control and a list of devices for realization of these methods. Selection of a control method should be based on knowledge about conditions of element's operation, its geometrical sizes, physical characteristics of element's material, suitability of elements for control and sensitivity of existing methods of control. Definition of parameter monitoring strategy consists in the formation of optimum schedule of estimation and analysis of safety of operation of technical systems by the given criterion.

The quality of solution to the problem of collecting of statistical and expert information and control by methods of non-destructing testing of parameters determines the quality of input data needed for estimation and analysis of safety.

The problem of risk estimation and analysis of current safety of operation of technical systems based on monitoring results supposes step-by-step fulfillment of stages of same algorithm.

The following problem of monitoring is the development of plan of actions on maintenance of demanded safety of operation of the technical system on the basis of results of estimation. Besides, in solving this problem, we can correct models of accident appearances, the list of controlled parameters and requirements to safe operation of GSI. At this, we update both the initiating data for models of accidents appearances at the GSI operation and the model's structure. The necessity of correction of the list of controlled parameters, determining safe operation of technical systems, is necessary in view of the changes of accident weights during monitoring. The correction of requirements to safety is made at each stage of monitoring.

The considered approach to the risk estimation and analysis of safe operation of elements of the GSI enables to estimate quantitatively the risk of safe operation with taking into account results of monitoring of technical condition parameters of objects to determine parameters, which are the most effective for the risk management, and to substantiate methods of safety management (in the framework of the concept of "acceptable risk").

The experience of usage of such system of monitoring for the space rocket systems "Proton" proves the high effectiveness as the information technology for solving problems of estimation and forecasting of technical condition and residual technical resources.

7.18 The Risk LP-Theory with GIE in the Classification Problem

The risk LP-theory with GIE in the classification problem will be considered, for the most part, by the example of classification of credits into *"good"* and *"bad"*. The considered approach allows us to construct a well-organized risk P-polynomial. As examples show, the accuracy of the risk LP-model is almost two times higher and robustness is almost seven times more, than those of methods based on the discriminant analysis and the neuron networks.

Methods of classification of credits. There are numerous papers and books about various methods of classification, risk measurement and econometrics (Goldenberg, 1990; Dougherty, 1992; Aivasyn and Mhitaryan, 1998;) based on the linear and quadratic discriminant analysis, the cluster analysis, the neuron networks and the theory of multidimensional normal distributions.

Classification of risk objects by methods of rates and recognitions. In the rating-based methods, the factors, taken into account, are the business risk (external environment, management quality, nature of mutual relation with clients, characteristics of credits, etc.) and the financial risk (factors of the company financial condition) (Solojentsev et al., 1999). Then, the risk rating system is used, which is the system of estimations of factors. By the value of the total parameter, the credit can be referred to one of two subclasses: bad or good objects (though, it is not clear how bad or good it is). The methods do not allow to find an exact numerical estimate of the risk, to establish the admitted risk, to determine the price for risk and to reveal parts of the system resulting in most failures.

In the methods, based on the use of mathematical classification methods, estimation of the credit risk is carried out using a set of data on the client and the credit. These data include $20 \div 38$ fields. A number of grades for each field are equal to $2 \div 11$. The separating surface attributes the credit to one of two subclasses: the good or bad credits. Here, the answer is not given to how good or bad the credit is. The quality of an object classification method is characterized by several parameters. If a class has only two subclasses, for example, good and bad objects, then such parameters are errors E_g, E_b in the classification of good and bad objects and the average error E_m in the classification of objects.

Classification of risk objects with the use of the neuron networks. In the classification methods based on the neuron networks (NN), the "objects and signs" table is used too. The NN is built by introducing an entrance layer of nodes, internal hidden layers of nodes and a final node. In the entrance layer, the number of nodes is equal to the number of signs in the object. The grades of signs come to each node from the recognized object. For the given structure, the training of NN consists in a choice of such weights of its edges that the maximal number of objects should be classified correctly (Solojentsev et al., 1999).

For any object *i*, weights of edges, which get to the final node, are summed up, and the weight C_i is calculated. In the final node, a stepped transfer function with the threshold C_{ad} is used for object classification to good and bad ones. The values $Ci,\ldots, CN$ and C_{ad} are not constrained any way. There are N_b bad (N_g, good) objects to the right (left) of the threshold C_{ad}. So, each object is at a known distance from C_{ad} and the objects can be compared by these distances.

If in NN-methods we also want to solve the problem of estimation of probability of non-success, it is necessary to introduce very rigid requirements to NN as shown below. These requirements actually mean that NN should work in the probabilistic space with the logic, instead of arithmetic, addition of events (weights of edges).

Logic and probabilistic estimation and risk analysis in engineering. The LP-method (Ryabinin, 1976, 2000) is applied to estimate the risk in complex technical systems. It is based on the logic representation of development of dangerous conditions and mathematical methods of calculation of the truth functions of logic algebra. The fundamental concept of the LP-theory is the concept of the dangerous condition and the system danger appropriate function. The risk structural model (scenario) is written. It represents the graph, and which nodes are connected by connection of types AND, OR, NOT. Each graph node can accept value 1 or 0. Some nodes of the graph are random events with known probabilities (initiating events); other nodes are derivative events. The probabilities of initiating events are known. The probabilities of derivative events are calculated.

The risk logic function (L-function) is made according to the graph with the help of the risk shortest ways, or with the help of the minimal cross-sections of risk prevention. We obtain the risk P-function after orthogonalization of the risk L-function. The risk computation is made with the risk P-model by substituting in it the appropriate probabilities of initiating events. The initiating events probabilities are assumed to be known either from tests of separate elements of the system or from experts' estimates.

The LP-method allows us to estimate numerically the object risk, to classify objects by the risk value and to analyze contributions of the initiating events to the object risk (Ryabinin, 1976, 2000). However, the direct application of the LP-method used in engineering is impossible for the estimation of the non-success risk in banks, business and quality of companies or production. In the development of the known LP-theory of reliability and safety by I. Ryabinin (Ryabinin, 1981, 1976, 2000), used in engineering, we introduce new concepts and risk problems (Solojentsev et al., 1999; Solojentsev and Karassev, 2003):

- initiating and final events are considered on many levels;
- associative risk LP-models constructed by using common sense for events connection are considered too;
- problems of parametrical and structural identification of risk LP-models from the statistical data are solved;
- new problems of the risk analysis on the basis of calculation of contributions of initiating events in the mean risk of the object set and in the accuracy of the risk LP-model are solved;
- the scenario management by the non-success risk at stages of designing, debugging and operational tests, and operation are considered.

7.19 Tabular Representation of Statistical Data

The statistical data for risk analysis and evaluation are presented in tabular form (Table 7.2). Table rows represent objects or states of the object $i = 1, 2, \ldots, N$. Table columns correspond to the object signs $Z_1, \ldots, Z_j, \ldots, Z_n$. The parameter in its turn can have grades Z_{rj}, $r = 1, 2, \ldots, N_j$. The grades are in cells of the table. The last table column contains the parameter of object effectiveness Y. The values of grades are considered as random variables or grade-events, which distribution is set by a discrete series, that is, by values of grades and their probabilities.

For measurement of signs or characteristics of the object, scales are used: logical (true or false, 1 or 0), qualitative/enumeration (blue, red, etc.), linear order $(a_1 > a_2 > \ldots > a_n)$, numerical (intervals $[a,b]$), etc.

Generally, the grades are not ordered linearly and one cannot say that grade 3 is worse or better than grade 4 for the final event. The final event also has grades. For example, a credit has the following grades: 1, if the credit is returned; 0, if the credit is not returned.

The signs and grades are corresponded to random events, which result in failures [7-g], and logical variables with the same symbols of identifiers. Sign-events (their number is equal to n) are connected logically by operations OR, AND, NOT. The grade-events for each sign form GIE.

Table 7.2 Objects and parameters

Objects i	*Sign* Z_1	...	*Sign* Z_j	...	*Sign* Z_n	*Output Parameter,* Y
1						
...	...	...	...	...	...	...
			Z_{jr}			
...	...	...	...	...	...	...
N						

7.19.1 Basic Equations

The binary logic variable Z_j is equal to 1 with probability P_j, if the sign j leads to the non-success; otherwise, Z_j is equal to 0 with probability $Q_j = 1-P_j$. The binary logic variable Z_{jr}, corresponding to the grade r of the sign j, is equal to 1 with probability P_{jr}; otherwise, it is equal to 0 with probability $Q_{jr} = 1-P_j$. Binary vector $Z(i)$ $(Z_1,\ldots, Zj,\ldots, Z_n)$ describes the object i from the table "Objects and signs". In assigning the object i instead of the logic variables $Z_1,\ldots, Z_j,\ldots, Z_n$, it is necessary to substitute the logic variable Z_{jr} for grades of signs for the object i. We write down the general form of the non-success risk L-function for any object

$$\Upsilon = \Upsilon\left(Z_1,\ldots,Z_j,\ldots,Z_n\right) \tag{7.22}$$

and the non-success risk P-function of any object, given by the vector $Z(i)$,

$$P_i\left\{\Upsilon = 1 \middle| Z(i) = \psi\left(P_1,\ldots,P_j,\ldots,P_n\right),\quad i = 1,2,\ldots,N\right\} \tag{7.23}$$

For each grade-event in GIE, we consider three probabilities: W_{jr} is the relative frequency of the grade in objects of the table "Objects and signs"; Pl_{jr} is the probability of the grade-event in GIE; P_{jr} is the probability of the grade-event to be substituted into instead of the probability P_j. We define these probabilities for the jth GIE as follows:

$$W_{jr} = P\left\{Z_{jr} = 1\right\};\quad \sum_{r=1}^{N} W_{jr} = 1;\quad r = 1,2,\ldots,N_j; \tag{7.24}$$

$$W_{jr} = P\left\{Z_{jr} = 1\right\};\quad \sum_{r=1}^{N} W_{jr} = 1;\quad r = 1,2,\ldots,N_j;$$

$$Pl_{jr} = P\left\{Z_{jr} = 1 \middle| Z_j = 1\right\};\quad \sum_{r=1}^{N} Pl_{jr} = 1;\quad r = 1,2,\ldots,N_j;$$

$$P_{jr} = \left\{Z_j = 1 \middle| Z_{jr} = 1\right\};\quad r = 1,2,\ldots,N_j$$

$$P_{jr} = \left\{Z_j = 1 \middle| Z_{jr} = 1\right\};\quad r = 1,2,\ldots,N_j$$

Here and subsequently, we shall use $j = 1,2,\ldots, n$; $r = 1, 2,\ldots, N_j$ for formulae, where n is the number of sign-events, and N_j is the number of grade-events in jth GIE.

The mean probabilities W_{jr}, Pl_{jr} and P_{jr} in GIE are equal to:

$$W_{jm} = 1/N_j;\ P_{jm} = \sum_{r=1}^{N} P_{jr}\cdot W_{jr};\ Pl_{jm} = \sum_{r=1}^{N} Pl_{jr}\cdot W_{jr}. \tag{7.25}$$

The object risk P_i is calculated by replacing probabilities P_{jr} by P_j. We shall estimate probabilities P_{jr} during the process of algorithmic iterative training (identification) of the risk P-model by using the data from the table "Objects and signs". In the beginning, it is necessary to determine the probabilities Pl_{jr}, satisfying, and further to pass from the probabilities Pl_{jr} to the probabilities P_{jr}. The number of the estimated independent probabilities P_{jr}, is equal to:

$$N_{\text{ind}} = \sum_{j=1}^{n} N_j - n. \tag{7.26}$$

The connection of the probabilities P_{jr}, and Pl_{jr} for the grades is expressed through the mean values of the probabilities P_{jm} and Pl_{jm}:

$$P_{jr} = Pl_{jr} \cdot \left(P_{jm} / Pl_{jm}\right); \quad r = 1, \ldots, N_j; \quad j = 1, \ldots, n. \tag{7.27}$$

7.19.2 Examples of Structural, Logic and Probabilistic Risk Models

The structural risk model can be equivalent to a realistic one (for example, electrical system); it can be associative if based on the common sense, or mixed. The non-success risk L-model of the "node" type (Figure 7.10(a)) is stated as follows:

$$Y = Z_1 \vee Z_2 \vee \ldots \vee Z_j \ldots \vee Z_n. \tag{7.28}$$

In words, it means that the non-success occurs if any one, or two,..., or all initiating events occur. After orthogonalization of the L-function, we have the following non-success risk P-model:

$$P = P_1 + P_2 \cdot Q_1 + P_3 \cdot Q_1 \cdot Q_2 + \cdots \tag{7.29}$$

In the risk LP-model, the "arithmetic's" is such that for the final event the risk value belongs to [0,1] for any values of probabilities of initiating events. The non-success risk L-model of the "bridge" type (Figure 7.10(b)) is represented in the normal disjunctive form as a logic sum of the shortest paths of successful operation:

$$Y = Z_1Z_3 \vee Z_2Z_4 \vee Z_1Z_5Z_4 \vee Z_2Z_5Z_3. \tag{7.30}$$

Orthogonalization provides the non-success risk P-model:

$$P_i = p_2p_4 + p_1p_3 + q_1p_2p_3p_4p_5 + p_1q_2q_3p_4p_5 - p_1p_2p_3p_4. \tag{7.31}$$

7.19.3 Measure and Cost of Risk

Let us introduce an admitted risk P_{ad} separating the objects into good and bad: if $P_i > P_{ad}$, then the object is bad; if $P_i < P_{ad}$, then the object is good (Figure 7.10(a)).

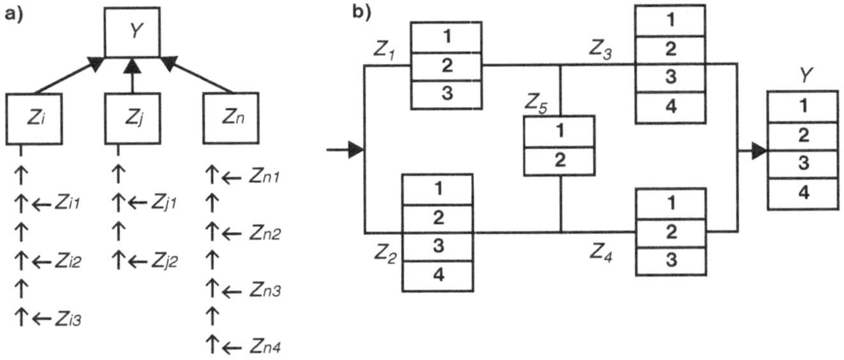

Figure 7.10 Structural models of risk: (a) unit type; (b) bridge type.

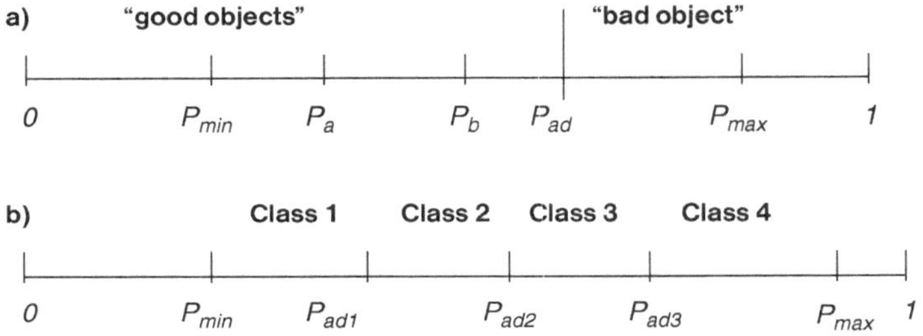

Figure 7.11 The scheme of classification of objects by risk: (a) into two classes; (b) into several classes.

If the objects are classified into a greater number of classes, then a corresponding number of admitted risks: P_{ad1}, P_{ad2},..., is introduced (Figure 7.10(b)).

Let us assume that the probabilities of grade-events P_{jr}, $j = 1, 2,..., n$; $r = 1,2,..., N_j$ are known. Then, from the risk P-model, we calculate risks of all N objects of the table "Objects and signs". We plot these risks on the risk axis. If the resulting event Y has two grades (Figure 7.10(a)), we choose the admitted risk P_{ad}, so that N_b from N objects are bad and N_g are good. For the object i, the distance between the risk P_i and the admitted risk P_{ad} is a natural measure of its being "good" or "bad":

$$d_i = |P_i - P_{ad}|. \tag{7.32}$$

The object risks can be represented in a different manner. We calculate the number of objects N_{ad} and N_i having risks, respectively, smaller than the admitted risk P_{ad} and smaller than the risk P_i of the object i and establish the following risk measures:

1 The relative number of objects having risks, respectively, smaller than a_i and greater than the risk b_i of the object i under consideration:

$$a_i = N_i / N;\ b_i = 1 - a_i; \tag{7.33}$$

2 The relative numbers of good, f_i, and bad, e_i, objects having risks greater than that of the considered object i among the good and bad objects:

$$f_i = (N_{ad} - N_i)/N_{ad};\quad e_i = (N_i - N_{ad})/(N - N_{ad}). \tag{7.34}$$

3 The aforementioned measures are used to calculate the cost of risk, for example, rate on credit. The simplest formula of the risk cost is as follows:

$$Ci = C_{ad} + C \cdot (P_i - P_{ad}), \tag{7.35}$$

where the cost of admitted risk C_{ad} and the coefficient C are chosen by the bank on the basis of the market conditions.

7.20 GIE and the Bayes Formula

Connection between probabilities of grades P_{jr} and Pl_{jm} in GIE is expressed in terms of the mean values of probabilities P_{jm} and Pl_{jm}. We shall prove that this fact follows from the Bayes formula. The condition probability $P(H_k/A)$ that a hypothesis H_k is true after the event A happens is given by the following formula:

$$P(H_k/A) = P(H_k) \cdot P(A/H_k)/P(A), \tag{7.36}$$

where

$$P(A) = \sum_{i=1}^{m} P(H_i) \cdot P(A/H_i), \tag{7.37}$$

and hypothesis H_i, $i = 1,\ldots, k,\ldots, m$, forms a complete GIE.

There are usually many GIE in risk problems. For X_j, each group forms a complete GIE of X_{jr}, $r = 1,\ldots, N_j$. Therefore, for simplicity, the following notation is introduced for the jth GIE that is equivalent to:

Event A = Sign-Event Z_j
Hypothesis H_k = Grade-Event Z_{jr}
Probability $P(H_k)$= W_{jr}
Probability $P(H_k/A)$ = Pl_{jr}
Probability $P(A/H_k)$ = P_{jr}
Probability $P(A)$ = P_{jm}

We are going to use the Bayes formula only for training the risk LP-model on the statistical data by solving the corresponding problem of optimization. Therefore, there is no sense in discussing here "a priori" and "a posteriori" probabilities in the real sense. The Bayes formula can be written down formally in terms of Pl_{jr} instead of P_j, or, on the contrary, in terms of P_{jr} instead of Pl_{jr}. For the procedure of optimization (identification) of the risk LP-model, the Bayes formula is written down in terms of probabilities P_{jr}:

$$P_{ir} = \left(Pl_{jr} \cdot P_{jm}\right) / W_{jr}, \tag{7.38}$$

which allows us to decrease by 1 the number of independent probabilities Pl_{jr} in the GIE in comparison with generation of the probabilities P_{jr}. Estimation of accuracy of the probabilities Pl_{jr} also becomes simpler – indeed, the sum of probabilities Pl_{jr} in GIE is equal to 1 (100%).

However, one meets difficulty in using because for a limited number of statistical data, the denominator can turn to zero. Therefore, it is suggested to make use of to relate the probabilities P_j, and Pl_{jr} in the GIE.

7.21 Dynamic Risk LP-Models

At the first sight, the risk LP-models are static models, because they do not include time explicitly. However, this opinion is erroneous, since actually risk LP-models are always used as dynamic models with substitution of probability values of initiating events at the real time. And the risk LP-models can be built from the initial stage as dynamic ones. Let us consider some ways of construction of dynamic risk LP-models.

As an example, we construct the risk LP-model for the classical figures of the Technical Analysis (TA) for the financial and commodity markets, represented by linear diagrams and Japanese candles (Figure 7.11) (Erlih, 1996).

For the classical figure "Head and Shoulders", we shall introduce the sign-events (axis X) and grade-events (axis Y), which are marked by asterisks (*) and have numbers. That is, signs and grades are used instead of values of rates (of currencies or goods) and discrete time. The total number of sign-events is equal to 7, and the total number of grade-events is equal to 26. Now, it is possible to distinguish a lot of figures "Head and Shoulders", which differ by their grades, and to calculate their risks.

For training the risk P-model, it is necessary to collect statistics. For this purpose, we look through the linear diagrams of the currency rate, choose classical figures of one type, fix signs, grades and the result of the success of the event of buying/selling ($Y = 1/0$). After the sufficient information is gathered, the risk model is trained and is used for forecasting risk of buying/selling for each new classical figure of this type.

The decisions in the risk P-model are made in the following sequence: (1) recognition of the classical figure in the dynamic mode from the diagram; (2) the decision making for buying/selling in dependence on the risk. It is also easy to

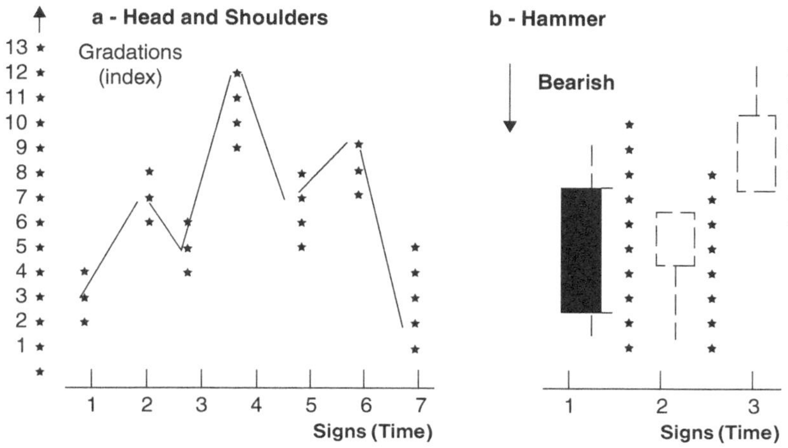

Figure 7.12 Construction of the dynamic risk LP-model.

take into account the factors of "Volume of sales" and "Open interest". For this purpose, we shall introduce two signs 8 and 9 (Figure 7.12). Let each of these signs have three grades. For example, sign 8 has grades: 1 means that the sale volume grows; 2 means that the sale volume does not change; 3 means that the sale volume falls. Thus, we construct the risk L-function by nine signs. Such model is hybrid, since it contains both sign-events (time) and usual sign-events.

In TA, the risk LP-model can be constructed for the Japanese candles (Figure 7.10) too. On the axis X, we introduce sign-events Z_1, Z_2, Z_3 for each of three candles in Figure 7.10. Besides, we introduce element events Z_{ll}, Z_{12}, Z_{13} for the description of the form of the first candle and other candles (top shadow, body and bottom shadow), respectively. For each element event, we shall define some event-grades (for example, Z_{121}, Z_{122}, Z_{123}, Z_{124}) for the body of the first candle, which are designated by asterisks * and have appropriate numbers on the axis Υ.

Now, after training of the risk LP-model, it is possible to distinguish many figures of different types and to calculate the risk for them.

7.22 Analysis of Results and Discussion with Proposed Taxonomy

We explore the literature and review the selected papers using the above-discussed taxonomy. To develop a holistic view of SCRM efforts, we included studies in practically all key demographical regions including Europe, Asia and the US. A combination of qualitative and quantitative approaches is adopted to describe the SCRM issues in the literature. The qualitative contents of the papers are provided in tables showing the issues discussed in the paper and also the approach adopted to address them.

We first review the papers focusing on the nature of the study and approach adopted. We found that an ample amount of work has been done but still it seems to be in a nascent state due to the paucity of normative studies. It is noted that more research initiatives have been taken with a positive approach (91%) than normative research. The low proportion of normative research (9%) exhibits the under-preparedness of research attempts to proffer precise and specific prescriptions to industries and academia.

Interestingly, we found that even after the decade-long period, the contribution of conceptual research is the highest, about 39%, followed by empirical 26% and analytical 35%. This finding suggests that the field of SCRM is still emerging and requires theoretical support to develop practical frameworks. Analytical approaches have also made a major contribution to assessing and characterizing the risk issues. But the feeble acceptance of these models in actual practices points out the need for more empirical studies to explore the critical underpinning elements and relationships of the risk appetite of firms, their propensity and financial status.

To provide the finer details, conceptual papers are further classified and it is observed that during 1996–2001, most of the papers focused on theoretical aspects related to risk issues, usually inspired by financial risk theories. But later on, catastrophic incidents such as the earthquake in Taiwan (2000), which severely damaged the supply base of the semiconductor industry; the Tsunami in Asia in 2005 that caused losses of more than $17 billion; Hurricane Katrina, which destroyed ports, railways, highways and communication networks and led to a significant drop in the US economy in 2006; terrorist attacks in the US and many Asian and European countries and many more motivated the researchers to redefine the risk issues for business continuity and devise mechanisms for quick recovery after disruptions. Thus, agility, resilience and flexibility in supply chains have become the core agenda for research. This has increased the contribution to the applied theory of SCRM, dealing with contemporary and upcoming issues.

Results show that theory is enhancing rapidly in the field of SCRM. Researchers are forming deeper insights and delving into critical SCRM aspects. Analysis also indicates that the field of SCRM is expanding but the attempts are still very small to review the prevalent literature. Thus, more reviews are required to unify the various research efforts and explore the latent dimensions of risk management to support the global SCRM efforts significantly.

We include the papers that used empirical approaches with surveys followed by statistical designs and structured case studies. Many papers are also noted that have a combination of both methods for quantitative and qualitative analyses. The empirical approaches have been used to establish the relationships among latent supply chain issues such as short supplies, supplier characteristics, demand variability, erratic behavior of customers and risk propensity (Blackhurst et al., 2005; Shokley and Ellis, 2006; Bailey and Francis, 2008). These methods refine the level of understanding of risks, which further helps in taking strategic and operational decisions (Devaraj et al., 2007; Sanders, 2008). It is recognized that survey-based statistical designs are the most adopted approach in empirical studies (52%) to develop the relationship models. But in the SCRM literature, case studies also have increasing acceptability to develop more specific qualitative and quantitative models.

In order to plan and coordinate in a risk environment, quantification of risk and analytical modeling is required. Based on the modeling approach, we categorize the literature into mathematical, simulation and agent-based methods for a variety of settings such as linear, integer, dynamic and stochastic.

The simple analytical approach to quantify and rank the risks is the analytical hierarchy process (AHP) with linear problem settings in a multi-attribute decision model. It reduces the complex decision problem into a series of one-to-one comparison followed by synthesis of results based on a hierarchical structure (Korpela et al., 2002; Gaudenzi and Borghesi, 2006; Levary, 2008). However, the subjectivity involved in AHP has always been a matter of concern.

Owing to the very nature of the risk, the stochastic models are more accepted in supply chains to model risk issues, varying from strategic to operational levels (Beamon, 1998). The uncertainty associated with variables is tackled mainly with three approaches. First, standard distributions are used in which continuous probability distributions are assigned for decision variables. Second, when continuous distribution is not feasible, discrete finite scenarios are established considering various combinations of uncertain parameters. Third, there are fuzzy approaches, where uncertainties in decision parameters are considered as fuzzy numbers and membership functions (Chen and Lee, 2004; Mele et al., 2007). Underlying complexities and impractical assumptions limit the utility of mathematical modeling. Moreover, in some cases, the explicit relationships between decision variables are difficult to model. In such situations, simulation techniques provide an alternative approach to analyzing the supply chains by constructing an artificial environment within which the dynamic behavior of the risks can be assessed. Various risk mitigation strategies and trade-offs are tested in a simulated environment with seasonality, level of information sharing, service level, net profit, etc., as simulation parameters (Labeau et al., 2000; Jammernegg and Reiner, 2007; Sohn and Lim, 2008; Thomas and David, 2008).

The simulation models also have certain limitations, such as the models can only be run with previously defined conditions and there are limited capabilities to design the system parameter itself (Swaminathan et al., 1998; Ohbyung et al., 2007). To overcome these shortcomings, multi-agent approaches, supported by advanced computational methods, have been introduced. In these approaches, the problem is modeled as agent elements (supplier, manufacturer, distributor, etc.), control elements (inventory control, scheduling, logistics and transportation, etc.) and their interaction protocols (Swaminathan et al., 1998; Mele et al., 2007). These approaches are better than individual programs as they combine the various autonomous agents/programs in one platform. Various strategic and operational issues such as collaboration under demand and supply uncertainties, the role of information sharing, inventory levels and robust and optimal designs are investigated and managerial inferences are drawn by researchers (Ohbyung et al., 2007; Mele et al., 2007; Chatzidimitriou et al., 2008).

7.23 Review Findings, Exploration of Gaps and Avenues for Future Research

SCRM is an exponentially growing area of research encompassing multidisciplinary and multidimensional aspects of risks. As the body of SCRM literature involves complex and entwined issues, a systematic taxonomy could make a great contribution. To delve into the supply chain risk issues, we presented a multi-layered top-down classification scheme. In the first layer, we considered the research approach and exploration of risk issues; in the second layer, we examined the nature of the study, research methods, orientation of risk definitions, structural elements and the level of implementation; in the third layer, the key discriminating elements of each factor were considered and were further categorized into detailed attributes. Apart from this, we have used a logical codification scheme employing an alphanumeric code which can assist in quantitative and qualitative analyses. We have further explored the literature with two very important and practical dimensions of the study, namely coordination and decision making in an uncertain business environment and implementation of SCRM for various sectors. The outcomes of these analyses have been presented in the form of propositions. In addition to describing the contributions of the researchers, this study also provided new insights for practical aspects of SCRM.

The conclusions of this chapter have illustrated the importance of adopting a broader view and scope of coordination strategies in the context of effective implementation of SCRM. It has been argued that understanding the emerging techniques, including conceptual, analytical and empirical approaches with all the proposed elements, enables us to tackle better the managerial challenges involved in addressing the risk issues. This kind of broader view is specifically needed in relation to the kind of managerial challenges faced by a company operating as a focal firm and having more power in supply chains. As this study has illustrated, it is not enough to concentrate on developing and sharpening the risk mitigation strategies focusing on one side of the supply chain and practices. Rather, the company needs to understand and try to influence the entire supply chain, or more importantly, the nature and progression of the flows across the various interfaces. The broadening of the scope of SCRM from a company's internal processes toward the inclusion of external issues is thus an important managerial challenge.

The review reveals various insights and gaps in the SCRM literature. On comparison of the nature of the study, it is observed that even though the literature has a plethora of work, the contribution of prescriptive studies is significantly lower, which justifies the need for more focused and specific studies, acceptable to industry. We noticed that the contribution of conceptual studies to SCRM has been higher than that of empirical and analytical studies. This finding highlights the fact that, as risk management studies are still in a nascent state, conceptual and theoretical upgrading is still essential to improve the level of understanding of complex risk issues to provide the strands of effective empirical and analytical studies. It has also been noted that SCRM is accepted in multiple research fields and the literature reflects a huge variety of works with diversified themes, issues and approach. The literature reports very few reviews covering the width and depth of the field. Moreover, as we found that the area is still emerging, more reviews are needed encompassing the changing trends in methodology, approach and finer elements of risk issues with various perspectives. Thus, attempts have been made in this study to cover the prevalent literature dealing with current research methods to address the risk issues.

The analysis of orientation of risk definitions suggests that operational aspects related to the demand supply mismatch and interruption of information, funds or material flow are the most utilized factors to define and classify risks. Market orientation factors such as customer expectations, market fluctuations, price variability, competitor moves, etc., are also found to be significant to characterize the risk issues. Strategic decision elements such as outsourcing, single sourcing, degree of leanness in manufacturing, level and type of coordination and information sharing, etc., are also issues of concern but are still not addressed as much as the operational elements. Moreover, product features such as life cycle, functionality and complexity in design have not been adequately explored to define the risk characteristics. Thus, including product and strategic perspectives to define the risks could improve the effectiveness of risk management mechanisms.

On exploration of the structural dimensions of the supply chain, it was observed that researchers emphasize supply side risks more than the demand side. The optimal number of suppliers, delivery reliability, optimal size of deliveries, relationships and coordination are the key elements that influence the risk management strategies, but in a changing scenario, customer-related element such as demand fluctuations and customer behavior should also be included to improve the agility and responsiveness of the supply chain. The implementation of a risk management program shows that scenario-based methods are more common due to their comprehensiveness to identify the risks, followed by listing methods due to their simplicity. Risk characterization techniques were found to be more accepted but are still not effective to quantify the elusive and dynamic nature of risks. Further, on investigation of risk management strategic stances, we found that the acceptor stance with the redesign of supply networks is more common than hitting the cause of risk and reshaping the uncertainty sources. After a series of natural and manmade disruptive events, recovery strategies are also being developed with the prime notion of robustness and resilience.

It has been noted that empirical studies primarily analyze the supply chain, investigating the impact of various risk factors on performance determinants, information sharing, collaboration and e-business practices. The implications of strategic moves such as outsourcing and lean practices have also been investigated with specific case studies and survey-based statistical analysis. However, as we know that the risk issues have strong perceptive elements and human and organizational behavior plays a decisive role in managing the risk situations, behavioral elements such as human/organization risk propensity can be integrated with the conventional risk models to get more realistic solutions. Moreover, the role of various personality traits, context and experience can also be incorporated in risk management models. Thus, empirical studies investigating behavioral and technical, as well as commercial aspects and their role in decision making will be more relevant to develop better risk management models.

The literature reflects the dramatic growth in mathematical modeling to analyze the risk issues. Initially, the problems were addressed with linear models but later on stochastic modeling and multi-agent approaches have been employed more to analyze the risk issues under simulated environments using artificial intelligence tools. To deal with supply chain risk issues, these models require further improvements. The literature reports various mathematical models developed to assist planning under uncertainties with a number of impractical assumptions such as known probability distributions and linearization in relationships, which reduce the acceptability of the model for real-life situations. Thus, inclusion of deeper risk issues can improve the effectiveness of mathematical models to a large extent.

It is also necessary to develop coordination strategies considering the actual conditions such as non-ideal members and heterogeneous risk sharing attitudes. Many times, managers have to analyze trade-offs considering the factors which contradict each other such as redundancy and efficiency. Methods and mechanisms are still required to analyze these trade-offs in a dynamic business environment with a risk perspective.

We have unified the study and analyzed it for coordination strategies under different decision making environments and implementation issues of SCRM for various sectors. The coordination strategies have been studied with two decision making scenarios, namely centralized and decentralized systems. In a centralized decision making environment, the level of coordination and information sharing among various players is found to be better but it is also observed that the firms leading the supply chain have the tendency to transfer the risks to smaller players. However, in a decentralized decision making environment, coordination is found only at the inter-firm level, which causes conflicting risk perceptions and practices to manage them. Based on the discussion, it can be said that coordination among various partners and appropriate level of information sharing is essential to improve the overall effectiveness of risk management strategies. Study further reflects the fact that different industries and sectors have different business environments, opportunities and limitations; thus, a common risk management framework may not be effective that causes the need for specific SCRMs for diversified industries and sectors.

Thus by employing a detailed taxonomy, we have investigated the prevalent SCRM literature focusing on the research methods adopted and exploration of the risk issues from definition to implementation phases and specific industry needs and we believe that the trend of growing interest in the field of SCRM will continue and new avenues will open from the strategic to the operation level with the inclusion of new developments in technology, computing techniques and managerial concerns to effectively manage the risk issues.

Appendices

Appendix A

Table A.1 Comparison of MCDM techniques

Procedure	*Computer support*	*Simplicity*	*Type of information*
Dominance	No	Yes	Det., Card.
Maximin	No	Yes	Det., Card.
Maximax	No	Yes	Det., Card.
Conjunctive	No	Yes	Det., Card., Ord.
Disjunctive	No	Yes	Det., Card., Ord.
Lexicography	No	Yes	Det., Card., Ord.
SAW	Yes	Yes	Det., Card.
TOPSIS	No	No	Det., Card.
ELECTRE I	Yes	No	Det., Card., Ord.
ELECTRE II	Yes	No	Det., Card., Ord.
ELECTRE III	Yes	No	Det., Card.
AHP	Yes	No	Det., NDet., Card.
NAIADE	Yes	No	Fuz., Card., Ord.
EVAMIX	No	Yes	Det., Card., Ord.
MAVT	Yes	No	Det., Card.
UTA	Yes	No	NDet., Card.
MAUT	Yes	No	NDet., Card.
SMART	Yes	No	Det., Card.
ORESTE	Yes	No	Det., Ord.
PROMETHEE	Yes	No	Det., Card., Ord.
REGIME	Yes	No	Det., Ord.
PAMSSEM	Yes	No	Fuz., Ord., Card., NDet.

Table A.2 Comparison of MCDM techniques

Procedure	*Compensation*	*Preference relations*	*Decision problem*
Dominance	None	$\{P, I\}$	A
Maximin	None	$\{P, I\}$	α, γ
Maximax	None	$\{P, I\}$	α, γ
Conjunctive	None	$\{P, I\}$	Screening
Disjunctive	None	$\{P, I\}$	Screening
Lexicography	None	$\{P, I\}$	A
SAW	Total	$\{P, I\}$	α, γ
TOPSIS	Partial	$\{P, I\}$	α, γ
ELECTRE I	Partial	$\{S, R\}$	A
ELECTRE II	Partial	$\{S, R\}$	Υ
ELECTRE III	Partial	$\{S, R\}$	Υ
AHP	Partial	$\{P, Q, I\}$	α, γ
NAIADE	Partial	$\{S, R\}$	Υ
EVAMIX	Partial	$\{S, R\}$	Υ
MAVT	Partial	$\{P, I\}$	α, γ
UTA	Partial	$\{P, I\}$	α, γ
MAUT	Partial	$\{P, I\}$	α, γ
SMART	Partial	$\{P, I\}$	α, γ
ORESTE	Partial	$\{P, I, R\}$	Υ
PROMETHEE	Partial	$\{P, I, R\}$	Υ
REGIME	Partial	$\{S, R\}$	Υ
PAMSSEM	Partial	$\{S, R\}$	α, γ

Appendix B

Table B.1 Main issues discussed over the years

Issues	*References*
1995~1999	
Financial risk management	Paradine (1995); Carr (1999)
Operation strategies	MacDuffie and Helper (1997); Williamson (1997); Yu (1997); Wilding (1998); Bowersox et al. (1999); Handfield et al. (1999)
	Bowersox, D.J., Stank, T.P. and Daugherty, P.J. Lean launch: managing product introduction risk through response-based logistics, *Journal of Product Innovation Management*, 1999. Vol. 16, pp. 557–568.
	Carr, N.G. Managing in the Euro zone. *Harvard Business Review*, 1999. Vol. 77, No. 1, pp. 47–48.
	Handfield, R.B., Ragatz, G.L., Petersen, K.J. and Monczka, R.M. Involving suppliers in new product development, *California Management Review*, 1999. Vol. 42, No. 1, pp. 59–82.
	MacDuffie, J.P. and Helper, S. Creating lean suppliers: diffusing lean production through the supply chain, *California Management Review*, 1997. Vol. 39, No. 4, pp. 118–151.
	Paradine, T.J. Business interruption insurance: a vital ingredient in your disaster recovery plan, *Information Management and Computer Security*, 1995. Vol. 3, No. 1, pp. 9–17.
	Wilding, R. The supply chain complexity triangle: uncertainty generation in the supply chain, *International Journal of Physical Distribution and Logistics Management*, 1998. Vol. 28, No. 8, pp. 599–616.
	Williamson, P.J. Asia's new competitive game, *Harvard Business Review*, 1997. Vol. 75, No. 5, pp. 55–67.
	Yu, G. Robust economic order quantity models, *European Journal of Operational Research*, 1997. Vol. 100, pp. 482–493.
2000~2004	
Environmental	Barry (2004); Cousins et al. (2004)
Financial risk management	LaLonde (2000; cited in Raphael, 2009); Cachon (2004); van Putten and MacMillan (2004)
Information management	Reichheld and Schefter (2000); Bradley (2001); Geary et al. (2002); Christopher and Lee (2004); Finch (2004); Lee (2004)

(*Continued*)

Issues	*References*
Operation strategies	Abernathy et al. (2000); Johnson (2001); Li et al. (2001); Raman et al. (2001); Tummala and Mak (2001); Svensson (2002); van der Vorst and Beulens (2002); Van Landeghem and Vanmaele (2002); Harland et al. (2003); Hauser (2003); Morales and Geary (2003); Peck et al. (2003); Rice and Caniato (2003); Berger et al. (2004); Chopra and Sodhi (2004); Christopher and Peck (2004); Svensson (2004)
Political and cultural practices	Giermanski (2000); Checa et al. (2003); Cousins et al. (2004)
Supply chain partners relationship	Svensson (2001); Hallikas et al. (2002); LaLonde (2002); Lee (2002); Zsidisin and Ellram (2003); Agrell et al. (2004); Hazra et al. (2004); Kamrad and Siddique (2004); Narayanan and Raman (2004)

Abernathy, F.H., Dunlop, J.T., Hammond, J.H. and Weil, D. Control your inventory in a world of lean retailing, *Harvard Business Review*, 2000. Vol. 78, No. 6, pp. 169–176.

Agrell, P.J., Lindroth, R. and Norrman, A. Risk, information and incentives in telecom supply chains, *International Journal of Production Economics*, 2004. Vol. 90, pp. 1–16.

Barry, J. Supply chain risk in an uncertain global supply chain environment, *International Journal of Physical Distribution and Logistics Management*, 2004. Vol. 34, No. 9, pp. 695–697.

Berger, P.D., Gerstenfeld, A. and Zeng, A.Z. How many suppliers are best? A decision-analysis approach, *Omega*, 2004. Vol. 32, pp. 9–15.

Bradley, P. The certainty of uncertainty, *Supply Chain Management Review*, 2001. Vol. 5, No. 2, pp. 105–106.

Cachon, G.P. The allocation of inventory risk in a supply chain: push, pull, and advance-purchase discount contracts, *Management Science*, 2004. Vol. 50, No. 2, pp. 222–238.

Checa, N., Maguire, J. and Barney, J. The new world disorders, *Harvard Business Review*, 2003. Vol. 81, No. 8, pp. 70–79.

Chopra, S. and Sodhi, M.S. Managing risk to avoid supply-chain breakdown, *MIT Sloan Management Review*, 2004. Vol. 46, No. 1, 53–62.

Christopher, M. and Lee, H. Mitigating supply chain risk through improved confidence, *International Journal of Physical Distribution and Logistics Management*, 2004. Vol. 34, No. 5, pp. 388–396.

Christopher, M. and Peck, H. Building the resilient supply chain, *International Journal of Logistics Management*, 2004. Vol. 15, No. 2, pp. 1–13.

Cousins, P.D. Lamming, R.C. and Bowen, F. The role of risk in environment-related supplier initiatives, *International Journal of Operations and Production Management*, 2004. Vol. 24, No. 6, pp. 554–565.

Finch, P. Supply chain risk management, *Supply Chain Management: An International Journal*, 2004. Vol. 9, No. 2, pp. 183–196.

Geary, S., Childerhouse, P. and Towill, D. Uncertainty and the seamless supply chain, *Supply Chain Management Review*, 2002. Vol. 6, No. 4, pp. 52–61.

Giermanski, J.R. A "Black Hole" on the border, *Supply Chain Management Review*, July/August 2000, 17–20.

Issues	*References*

Hallikas, J., Virolainen, V.M. and Tuominen, M. Risk analysis and assessment in network environments: a dyadic case study, *International Journal of Production Economics*, 2002. Vol. 78, pp. 45–55.

Harland, C., Brenchley, R. and Walker, H. Risk in supply networks, *Journal of Purchasing and Supply Management*, 2003. Vol. 9, pp. 51–62.

Hauser, L.M. Risk-adjusted supply chain management, *Supply Chain Management Review*, 2003. Vol. 7, No. 6, pp. 64–71.

Hazra, J., Mahadevan, B. and Seshadri, S. Capacity allocation among multiple suppliers in an electronic market, *Production and Operations Management*, 2004. Vol. 13, No. 2, pp. 161–170.

Johnson, M.E. Learning from toys: lessons in managing supply chain risk from the toy industry, *California Management Review*, 2001. Vol. 43, No. 3, pp. 106–124.

Kamrad, B. and Siddique, A. Supply contracts, profit sharing, switching, and reaction options, *Management Science*, 2004. Vol. 50, No. 1, pp. 64–82.

LaLonde, B.J. Who can you trust these days? *Supply Chain Management Review*, May/June 2002, pp. 9–10.

Lee, H.L. Aligning supply chain strategies with product uncertainties, *California Management Review*, 2002. Vol. 44, No. 3, pp. 105–119.

Lee, H.L. The triple-A supply chain, *Harvard Business Review*, October 2004, 102–112.

Li, L., Porteus, E.L. and Zhang, H. Optimal operating policies for multiplane stochastic manufacturing systems in a changing environment, *Management Science*, 2001. Vol. 47, No. 11, pp. 1539–1551.

Morales, D.K. and Geary, S. Speed kills: supply chain lessons from the war in Iraq, *Harvard Business Review*, 2003. Vol. 81, No. 11, pp. 16–17.

Narayanan, V.G. and Raman, A. Aligning incentives in supply chains, *Harvard Business Review*, 2004. Vol. 82, No. 11, pp. 94–102.

Peck, H., Abley, J., Christopher, M., Haywood, M., Saw, R., Rutherford, C. and Strathern, M. *Creating Resilient Supply Chains: A Practical Guide*. Cranfield University, Cranfield School of Management, UK. 2003.

Raman, A., DeHoratius, N. and Ton, Z. The Achilles' heel of supply chain management, *Harvard Business Review*, 2001. Vol. 79, No. 5, pp. 25–28.

Reichheld, F.F. and Schefter, P. E-loyalty: your secret weapon on the web, *Harvard Business Review*, July/August 2000, pp. 105–113.

Rice, J.B. and Caniato, F. Building a secure and resilient supply network, *Supply Chain Management Review*, 2003. Vol. 7, No. 5, pp. 22–30.

Svensson, G. A conceptual framework of vulnerability in firms' inbound and outbound logistics flows, *International Journal of Physical Distribution and Logistics Management*, 2002. Vol. 32, No. 2, pp. 110–134.

(*Continued*)

Issues	*References*
	Svensson, G. Key areas, causes and contingency planning of corporate vulnerability in supply chains: a qualitative approach, *International Journal of Physical Distribution and Logistics Management*, 2004. Vol. 34, No. 9, pp. 728–748.
	Svensson, G. Perceived trust towards suppliers and customers in supply chains of the Swedish automotive industry, *International Journal of Physical Distribution and Logistics Management*, 2001. Vol. 31, No. 9, pp. 647–662.
	Tummala, V.M.R. and Mak, C.L. A risk management model for improving operation and maintenance activities in electricity transmission networks, *Journal of the Operational Research Society*, 2001. Vol. 52, No. 2, pp. 125–134.
	van der Vorst, J.G.A.J. and Beulens, A.J.M. Identifying sources of uncertainty to generate supply chain redesign strategies, *International Journal of Physical Distribution and Logistics Management*, 2002. Vol. 32, No. 6, pp. 409–430.
	Van Landeghem, H. and VanMaele, H. Robust planning: a new paradigm for demand chain planning, *Journal of Operations Management*, 2002. Vol. 20, pp. 769–783.
	van Putten, A.B. and MacMillan, I.C. Making real options really work, *Harvard Business Review*, 2004. Vol. 82, No. 12, pp. 134–141.
	Zsidisin, G.A. Managerial perceptions of supply risk, *Journal of Supply Chain Management*, 2003. Vol. 39, pp. 14–26.
2005~2009	
Environmental	LaLonde (2005); Peck (2005); Stalk (2006); Economy and Lieberthal (2007); Murphy (2007)
Financial risk management	Callioni et al. (2005); Hendricks and Singhal (2005); Bovet (2006); Choi and Krause (2006); Hartley-Urquhart (2006); Kerr (2006); Papadakis (2006); Fang Hillman and Whinston (2007); Smith et al. (2007); Xiao and Qi (2008)
Information management	Jharkharia and Shankar (2005); Faisal et al. (2006, 2007); Ratnasingam (2007)
Operation strategies	Elkins et al. (2005); Hale and Moberg (2005); Jüttner (2005); Sheffi and Rice (2005); Zsidisin and Smith (2005); Zsidisin et al. (2005); Gattorna (2006); Kiser and Cantrell (2006); Khan et al. (2008)
Outsourcing to low cost countries	Fitzgerald (2005); Amaral et al. (2006); Cigolini and Rossi (2006); Crone (2006); Kumar et al. (2007)
Political	Kleindorfer and Saad (2005); de Waart (2006); Stalk (2006); Economy and Lieberthal (2007); Ferrer et al. (2007); Cudahy et al. (2008)
Supply chain partners relationship	Martínez-de-Albéniz and Simchi-Levi (2005); Levary (2007)
	Amaral, J., Billington, C.A. and Tsay, A.A. Safeguarding the promise of production outsourcing, *Interfaces*, 2006. Vol. 36, No. 3, pp. 220–233.
	Bovet, D. The self-funding supply chain, *Supply Chain Management Review*, 2006. Vol. 10, No. 5, pp. 9–10.
	Callioni, G., de Montgros, X., Slagmulder, R., Van Wassenhove, L.N. and Wright, L. Inventory-driven costs, *Harvard Business Review*, 2005. Vol. 83, No. 3, pp. 135–141.
	Choi, T.Y. and Krause, D.R. The supply base and its complexity: implications for transaction costs, risks, responsiveness, and innovation, *Journal of Operations Management*, 2006. Vol. 24, pp. 637–652.

Issues	*References*
	Cigolini, R. and Rossi, T. A note on supply risk and inventory outsourcing, *Production Planning and Control*, 2006. Vol. 17, No. 4, pp. 424–437. Crone, M. Are global supply chains too risky? A practitioner's perspective, *Supply Chain Management Review*, 2006. Vol. 10, No. 4, pp. 28–35. Cudahy, G., Mulani, N. and Cases, C. Mastering global operations in multipolar world, *Supply Chain Management Review*, 2008. Vol. 12, No. 2, pp. 22–29. de Waart, D. Getting smart about risk management, *Supply Chain Management Review*, 2006. Vol. 10, No. 8, pp. 27–33. Economy, E. and Lieberthal, K. Scorched earth, *Harvard Business Review*, 2007. Vol. 85, No. 6, pp. 88–96. Elkins, D., Handfield, R.B., Blackhurst, J. and Craighead, C.W. 18 ways to guard against disruption, *Supply Chain Management Review*, 2005. Vol. 9, No. 1, pp. 46–53. Faisal, M.N., Banwet, D.K. and Shankar, R. Information risks management in supply chains: an assessment and mitigation framework, *Journal of Enterprise Information Management*, 2007. Vol. 20, No. 6, pp. 677–699. Faisal, M.N., Banwet, D.K. and Shankar, R. Mapping supply chains on risk and customer sensitivity dimensions, *Industrial Management and Data Systems*, 2006. Vol. 106, No. 6, pp. 878–895. Fang, F. and Whinston, A. Option contracts and capacity management—enabling price discrimination under demand uncertainty, *Production and Operations Management*, 2007. Vol. 16, No. 1, pp. 125–137. Ferrer, J., Karlberg, J. and Hintlian, J. Integration: the key to global success, *Supply Chain Management Review*, 2007. Vol. 11, No. 2, pp. 24–30. Fitzgerald, K.R. Big savings, but lots of risk, *Supply Chain Management Review*, 2005. Vol. 9, No. 9, pp. 16–20. Gattorna, J. Supply chains are the business, *Supply Chain Management Review*, 2006. Vol. 10, No. 6, pp. 42–49. Hale, T. and Moberg, C.R. Improving supply chain disaster preparedness: a decision process for secure site location, *International Journal of Physical Distribution and Logistics Management*, 2005. Vol. 35, No. 3, pp. 195–207. Hartley-Urquhart, R. Managing the financial supply chain, *Supply Chain Management Review*, September 2006, 18–25. Hendricks, K.B. and Singhal, V.R. An empirical analysis of the effect of supply chain disruptions on long-run stock price performance and equity risk of the firm, *Production and Operations Management*, 2005. Vol. 14, pp. 35–52. Jharkharia, S. and Shankar, R. IT-enablement of supply chains: understanding the barriers, *The Journal of Enterprise Information Management*, 2005. Vol. 18, No. 1, pp. 11–27. Jüttner, U. Supply chain risk management: understanding the business requirements from a practitioner perspective, *The International Journal of Logistics Management*, 2005. Vol. 16, No. 1, pp. 120–141.

(*Continued*)

Issues	*References*
	Kerr, J. Streamlining the cash flow, *Supply Chain Management Review*, October 2006, S25–S31.
	Khan, O., Christopher, M. and Burnes, B. The impact of product design on supply chain risk: a case study, *International Journal of Physical Distribution and Logistics Management*, 2008. Vol. 38, No. 5, pp. 412–432.
	Kiser, J. and Cantrell, G. 6 steps to managing risk, *Supply Chain Management Review*, 2006. Vol. 10, No. 3, pp. 12–17.
	Kleindorfer, P.R. and Saad, G.H. Managing disruption risks in supply chains, *Production and Operations Management*, 2005. Vol. 14, No. 1, pp. 53–68.
	Kumar, S., DuFresne, C. and Hahler, K. Managing supply chain risks in US-China trade partnership, *Information Knowledge Systems Management*, 2007. Vol. 6, pp. 343–362.
	LaLonde, B.J. Time to get serious about energy, *Supply Chain Management Review*, May/June 2005, pp. 8–9.
	Levary, R.R. Ranking foreign suppliers based on supply risk, *Supply Chain Management: An International Journal*, 2007. Vol. 12, No. 6, pp. 392–394.
	Martínez-de-Albéniz, V. and Simchi-Levi, D. A portfolio approach to procurement contracts, *Production and Operations Management*, 2005. Vol. 14, No. 1, pp. 90–114.
	Murphy, S. The supply chain in 2008, *Supply Chain Management Review*, December 2007, pp. 4–7.
	Papadakis, I.S. Financial performance of supply chains after disruptions: an event study, *Supply Chain Management: An International Journal*, 2006. Vol. 11, No. 1, pp. 25–33.
	Ratnasingam, P. A risk-control framework for e-marketplace participation: the findings of seven cases, *Information Management and Computer Security*, 2007. Vol. 15, No. 2, pp. 149–166.
	Sheffi, Y. *The Resilient Enterprise: Overcoming Vulnerability for Competitive Advantage*. Cambridge, MA, MIT Press. 2005.
	Smith, G.E., Watson, K.J., Baker, W.H. and Pokorski, J.A. A critical balance: collaboration and security in the IT-enabled supply chain, *International Journal of Production Research*, 2007. Vol. 45, No. 11, pp. 2595–2613.
	Stalk, G. The costly secret of China sourcing, *Harvard Business Review*, 2006. Vol. 84, No. 2, pp. 64–66.
	Xiao, T. and Qi, X. Price competition, cost and demand disruptions and coordination of a supply chain with one manufacturer and two competing retailers, *Omega-The International Journal of Management Science*, 2008. Vol. 36, pp. 741–753.
	Zsidisin, G.A., Melnyk, S.A. and Ragatz, G.L. An institutional theory perspective of business continuity planning for purchasing and supply management, *International Journal of Production Research*, 2005. Vol. 43, No. 16, pp. 3401–3420.
	Zsidisin, G.A. and Smith, M.E. Managing supply risk with early supplier involvement: a case study and research propositions, *The Journal of Supply Chain Management: A Global Review of Purchasing and Supply Fall*, 2005. pp. 44–57.

References

Accidents and Catastrophes: Preventing and Estimation of Consequences. Four volumes, K.E. Kochetkova and V.A. Kotlyarovsltogo (eds.). Moscow: Association of building Institutes, 1995–1997.

Aivasyn, S.I. and Mhitaryan, V.S. *Applied Statistics and Basis of Econometrics.* Moscow: Uniti, 1998.

Akin, E. *The Geometry of Population Genetics.* Berlin: Springer-Verlag, 1979.

Akintoye, A.S. and MacLeod, M.J. Risk analysis and management in construction, *International Journal of Project Management*, 1997. Vol. 15, pp. 31–38.

Albrecht, W., Wernz, G. and Williams, T. *Fraud: Bringing Light to the Dark Side of Business.* Transl. from English. Saint Petersburg: Piter, 1995.

Analysis Statistical Methods of Safety of Complex Technical Systems, Textbook, V.P. Sokolov (ed.). Moscow: Logos, 2001.

Anderson, D.R., Sweeney, D.J., Williams, T.A., Freeman, J. and Shoesmith, E. *Statistics for Business and Economics*, Second edition. England: Thomson. 2007.

Anderson, P.W., Arrow, K.J. and Pines, D. (eds.). *The Economy as an Evolving Complex System.* Redwood City: Addison-Wesley, 1988.

Annikeichik, N.D., Devyatkin, A. M. and Krasnov, O.V. Analysis of fire models at accident of rocket launching pad, *Journal of VISA named A.F.Mojaisky*, 2000. Vol. 7, pp. 54–59.

Annikeichik, N.D. and Sokolov, E.I. *Development of method of monitoring of safety of operation of technical systems of rocket launching pad.* Proceedings of the First International Scientific School: Modelling and Analysis of Safety and Risk in Complex Systems; Saint Petersburg: Omega, 2001 June 18–22.

Arunachalam, V. and Jegadheesan, C. Modified failure mode and effects analysis: a reliability and cost-based approach, *The ICFAI Journal of Operations Management*, 2006. Vol. 5, No. 1, pp. 7–20.

Atkinson, R. Project management: cost, time and quality, two best guesses and a phenomenon, it's time to accept other success criteria, *International Journal of Project Management*, 1999. Vol. 17, No. 6, pp. 337–342.

Bailey, K. and Francis, M. Managing information flows for improved value chain performance, *International Journal of Production Economics*, 2008. Vol. 111, No. 1, pp. 2–12.

Barfod, M.B., Salling, K.B., and Leleur, S. Composite decision support by combining cost-benefit and multi-criteria decision analysis. *Decision Support Systems.* 2011. Vol. 51, 167–175.

Barkow, J.H., Cosmides, L. and Tooby, J. (eds.). *The Adapted Mind: Evolutionary Psychology and the Generation of Culture.* New York: Oxford University Press, 1992.

Barron, FH and Barrett, BE. The efficacy of SMARTER—Simple multi-attribute rating technique extended to ranking, *Acta Psychologica*, 1996. Vol. 93, No. 1–3, pp. 23–36.

Beamon, B.M. Measuring supply chain performance, *International Journal of Operations & Production Management*, 1999. Vol. 19, No. 3, pp. 275–292.

Beamon, B.M. and Ware, T.M. A process quality model for the analysis, improvement and control of supply chain systems, *International Journal of Physical Distribution & Logistics Management*, 1998. Vol. 28, No. 9/10, pp. 704–715.

Beinat, E. and Nijkamp, P. *Multicriteria Analysis for Land-Use Management.* Dordrecht: Kluwer Academic Publishers, 1998.

Belton V, Stewart T. *Multiple Criteria Decision Analysis: An Integrated Approach.* Berlin, Germany: Springer Science & Business Media, 2002.

Beresford, A., Pettit, S. and Liu, Y. Multimodal supply chains: iron ore from Australia to China, *Supply Chain Management: An International Journal*, 2011. Vol. 16, No. 1, pp. 32–42.

Betzig, L., Borgerhoff-Mulder, M. and Turke, P. (eds.). *Human Reproductive Behavior: A Darwinian Perspective.* Cambridge: Cambridge University Press, 1988.

Blackhurst, J., Craighead, C.W., Elkins, D. and Handfield, R.B. An empirically derived agenda of critical research issues for managing supply-chain disruptions, *International Journal of Production Research*, 2005. Vol. 43, No. 19, pp. 4067–4081.

Blackhurst, J., Wu, T. and O'grady, P. A network-based decision tool to model uncertainty in supply chain operations, *Production Planning and Control*, 2007. Vol. 18, No. 6, pp. 526–535.

Blome, C. and Schoenherr, T. Supply chain risk management in financial crises-a multiple case-study approach, *International Journal of Production Economics*, 2011. Vol. 134, No. 1, pp. 43–57.

Blos, M.F., Quaddus, M., Wee, H.M. and Watanabe, K. Supply chain risk management (SCRM): a case study on the automotive and electronic industries in Brazil, *Supply Chain Management: An International Journal*, 2009. Vol. 14, No. 4, pp. 247–252.

Bluvband Z, Grabov P. *Failure analysis of FMEA.* In 2009 Annual Reliability and Maintainability Symposium (pp. 344–347). IEEE, Jan 26, 2009.

Bogataj, D. and Bogataj, M. Measuring the supply chain risk and vulnerability in frequency space, *International Journal of Production Economics*, 2007. Vol. 108, No. 1–2, pp. 291–301.

Bonner, J.T. *The Evolution of Complexity by Means of Natural Selection.* Princeton: Princeton University Press, 1988.

Bourguignon, B. and Mossart, D.L. The ORESTE method for multicriteria decision making in experimental chemistry, *Chemometric and Intelligent Laboratory Systems*, 1994. Vol. 22, No. 2, pp. 241–256.

Boute, R.N., Disney, S.M., Lambrecht, M.R. and Houdt, B.V. An integrated production and inventory model to dampen upstream demand variability in the supply chain, *European Journal of Operational Research*, 2007. Vol. 178, No. 1, pp. 121–142.

Bowersox, D.J. *Readings in Physical Distribution Management: The Logistics of Marketing.* Bowersox, D.J., La Londe, B.J. and Smykay, E.W. (eds.). New York: MacMillan, 1969.

Bowles, S. *Microeconomics: Behavior, Institutions and Evolution.* New York: Russell Sage; Princeton: Princeton University Press, 2004.

Boyd, R. and Richerson, P.J. *Culture and the Evolutionary Process.* Chicago: University of Chicago Press, 1985.

Brandon, R.N. and Burian, R.M. *Genes, Organisms, Populations: Controversies over the Units of Selection*. Cambridge, MA: The MIT Press, 1984.

Brans, J.P. and Mareschal, B. PROMETHEE Methods. In Multiple Criteria Decision Analysis: State of the Art Surveys, pp. 163–186. New York: Springer, 2005.

Brindley, C. *Supply Chain Risk*. Aldershot: Ashgate Publishing, 2004.

Bryson, K., Millar, H., Joseph, A. and Mobolurin, A. Using formal MS/OR modeling to support disaster recovery planning, *European Journal of Operational Research*, 2002. Vol. 141, No. 3, pp. 679–688.

Burtonshaw-Gunn, S.A. *The Essential Management Toolbox: Tools, Models and Notes for Managers and Consultants*. New York: John Wiley & Sons, 2008.

Callaghan, A. and Kemper, L. *A 2-phase aspiration-level and utility theory approach to large scale design*. Master's thesis, State University of New York at Buffalo, 2000.

Callaghan, A. and Lewis, K. *A 2-phase aspiration-level and utility theory approach to large scale design*. Master's thesis, State University of New York at Buffalo, 2000.

Carr, M., Konda, S., Monarch, I., Ulrich, C. and Walker, C. *Taxonomy Based Risk Identification (CMU/SEI-93-TR-6, ADA266992)*. Pittsburgh, PA: Software Engineering Institute, Carnegie Mellon University, 1993.

Cassill, D. Skew selection: nature favors a trickle-down distribution of resources in ants, *Journal of Bioeconomics*, 2003. Vol. 5, No. 2&3, pp. 83–96.

Cavinato, J.L. Supply chain logistics risks: from the back room to the board room, *International Journal of Physical Distribution and Logistics Management*, 2004. Vol. 34, No. 5, pp. 383–387.

Charnes, A. and Cooper, W. *Management Models and Industrial Applications of Linear Programming*. New York: Wiley, 1961, Vol. 1.

Charnov, E.L. Optimal foraging, the marginal value theorem, *Theoretical Population Biology*, 1976. Vol. 9, pp. 129–136.

Chatzidimitriou, K.C., Syneonidis, A.L., Kontogounis, I. and Mitkas, P.A. Agent mertacor: a robust design for dealing with uncertainty and variation in SCM environments, *Expert System with Applications*, 2008. Vol. 35, No. 3, pp. 591–603.

Chaudhuri, A., Mohanty, B. and Singh, K. Supply chain risk assessment during new product development: a group decision making approach using numeric and linguistic data, *International Journal of Production Research*, 2013. Vol. 51, No. 10, pp. 2790–2804.

Chen, C.L. and Lee, W.C. Multi objective optimization of multi-echelon supply chain networks with uncertain demands and prices, *Computers and Chemical Engineering*, 2004. Vol. 28, No. 6–7, pp. 1131–1144.

Chen, J.K. Utility priority number evaluation for FMEA, *Journal of Failure Analysis and Prevention*, 2007. Vol. 7, No. 5, pp. 321–328.

Cheng, S.K. and Kam, B.H. A conceptual framework for analyzing risk in supply networks, *Journal of Enterprise Information Management*, 2008. Vol. 21, No. 4, pp. 345–360.

Chiong, R. *Nature-Inspired Algorithms for Optimization*, Series in Studies in Computational Intelligence. Berlin: Springer-Verlag, 2009, Vol. 193.

Chopra, S. and Sodhi, M.S. Managing risk to avoid supply–chain breakdown, *MIT Sloan Management Review*, 2004. Vol. 46, No. 1, pp. 53–62.

Christopher, M. and Lee, H. Mitigating supply chain risk through improved confidence, *International Journal of Physical Distribution and Logistics Management*, 2004. Vol. 34, No. 5, pp. 388–396.

Christopher, M., Mena, C., Khan, O. and Yurt, O. Approaches to managing global sourcing risk, *Supply Chain Management: An International Journal*, 2011. Vol. 16, No. 2, pp. 67–81.

Christopher, M. and Peck, H. Building the resilient supply chain, *International Journal of Logistics Management*, 2004. Vol. 15, No. 2, pp. 1–14.

Churchman, C.W., Ackoff, R.L. and Arnoff, E.L. *Introduction to Operations Research.* New York: Wiley, 1957.

Clímaco, J.C. and Dias, L.C. An approach to support negotiation processes with imprecise information multicriteria additive models. In: Workshop on formal and informal information exchange in negotiations. *Proceedings of School of Information Technology and Engineering, University of Ottawa.* 2005.

Colicchia, C. and Strozzi, F. Supply chain risk management: a new methodology for a systematic literature review, *Supply Chain Management: An International Journal*, 2012. Vol. 17, No. 4, pp. 403–418.

Cooper, M.C., Lambert, D.M. and Pagh, J.D. Supply chain management: more than a new name for logistics, *International Journal of Logistics Management*, 1997. Vol. 8, No. 1, pp. 1–13.

COSO, *Enterprise Risk Management-Integrated Framework, The Committee of Sponsoring Organizations of the Treadway Commission (COSO)*, executive summary report, 2004.

Dailun, S.H.I. A review of enterprise supply chain risk management, *Journal of System Science and System Engineering*, 2004. Vol. 3, No. 2, pp. 219–244.

Dawkins, R. *The Selfish Gene.* New York: Oxford University Press, 1976.

Dawkins, R. *The Extended Phenotype: The Long Reach of the Gene.* San Francisco, CA: W.H. Freeman, 1982.

Deep, A. and Dani, S. *Managing Global Food Supply Chain Risks: A Scenario Planning Perspective*, Production and Operations Management Society, 20th Annual Conference, Orlando, Florida, USA, 2009. May 1–4, p. 21.

Delhaye, C., Teghem, J. and Kunsch, P. Application of the ORESTE method to a nuclear waste management, *International Journal of Production Economics*, 1991. Vol. 24, No. 1–2, pp. 29–39.

Detrain, C., Deneubourg, J.L. and Pasteels, J.M. (eds.) *Information Processing in Social Insects.* Basel: Birkhäuser Verlag, 1999.

Devaraj, S., Krajewski, L. and Jerry, C.W. Impact of eBusiness technologies on operational performance: the role of production information integration in the supply chain, *Journal of Operations Management*, 2007. Vol. 25, No. 6, pp. 1199–1216.

Diabat, A., Govindan, K. and Panicker, V. Supply chain risk management and its mitigation in a food industry, *International Journal of Production Research*, 2012. Vol. 50, No. 11, pp. 3039–3050.

Dillon, J.L. and Perry, C. Maldistributed utility theory, multiple objectives, and uncertainty in ex-ante project evaluation, *Review of Marketing and Agricultural Economics*, 1977. Vol. 45, No. 1&2, pp. 3–27.

Dougherty, C. *Introduction to Econometrics.* New York: Oxford University Press, 1992.

Dupré, J. (ed.). *The Latest on the Best: Essays on Evolution and Optimality.* Cambridge, MA: The MIT Press, 1987.

Dziadosz, A. and Rejment, M. Risk analysis in construction project-chosen methods, *Procedia Engineering*, 2015. Vol. 122, pp. 258–265.

Ebrat, M. and Ghodsi, R. Construction project risk assessment by using adaptive-network-based fuzzy inference system: an empirical study. *KSCE Journal of Civil Engineering*, 2014. Vol. 18, pp. 1213–1227.

Endler, J.A. *Natural Selection in the Wild.* Princeton: Princeton University Press, 1986.

Erlih A. *Hand-book in Technical Analysis for Commodity and Financial Markets.* Moscow: Infra, 1996.

Fandel, G. and Spronk, J. (eds). (1985). *Multiple Criteria Decision Methods and Applications*. Selected readings of the First International Summer School Acireale, Sicily. Berlin. Springer, September 1983.

Fiala, P. Information sharing in supply chains, *The International Journal of Management Science*, 2005. Vol. 33, No. 3, pp. 419–423.

Fishburn, P. A note on recent developments in additive utility theories for multiple factors situations, *Operations Research*, 1966. Vol. 14, pp. 1143–1148.

Fishburn, P. Methods for estimating additive utilities, *Management Science*, 1967. Vol. 13, pp. 435–453.

Fishburn, P.C. A survey of maldistributed/multiple criteria evaluation theories. In S. Zionts (ed.) *Multiple Criteria Problem Solving*. Berlin: Springer, 1978, pp. 181–224.

Fishburn, P.C. Utility theory, management science, *INFORMS Journal*, 1968. Vol. 14, pp. 335–378.

Fishburn, P.C. *Utility Theory for Decision Making*. New York: John Wiley and Sons, 1970.

Forrester, J.W. *Industrial Dynamics*. Cambridge, MA: Productivity Press, 1961.

Forrester, J.W. System dynamics, systems thinking, and soft OR, *System Dynamics Review*, 1994. Vol. 10, No. 2–3, pp. 245–256.

Franceschini, F. and Galetto, M. A new approach for evaluation of risk priorities of failure modes in FMEA, *International Journal of Production Research*, 2001. Vol. 39, No. 13, pp. 2991–3002.

Franks, N.R., Mallon, E.B., Bray, H.E., Hamilton, M.J. and Mischler, T.C. Strategies for choosing between alternatives with different attributes: exemplified by house-hunting ants, *Animal Behaviour*, 2003. Vol. 65, pp. 215–223.

French, S. *Reading in Decision Analysis*. London: Chapman and Hall, 1988.

Frolov, K.V. and Bulatov, V.P. *Fundamental and applied researches in the area of the safety theory and dynamic systems*. Proceedings of the First International Scientific School: Modelling and Analysis of Safety and Risk in Complex Systems; 2001 June 18–22; Saint Petersburg: Omega, 2001.

Frolov, K. V. and Mahutov, N. A. Problems of safety of complex engineering systems. In Book: *Problems of Machine-Building and Reliability of machines*. Moscow: Nauka, 1992, pp. 3–11.

Fuller, J., O'Connor, J. and Rawlinson, R. Tailored logistics: the next advantage, *Harvard Business* Review, 1993. Vol. 71, pp. 87–93.

Munda, G. *Multicriteria Evaluation in a Fuzzy Environment*, Contributions to economics Series. Heidelberg: Physica-Verlag, 1995.

Gaudenzi, B. and Borghesi, A. Managing risks in supply chain using AHP method, *The International Journal of Logistics Management*, 2006. Vol. 17, No. 1, pp. 114–136.

Gerber, M. and von Solms, R. Management of risk in the information age, *Computer and security*, 2005. Vol. 24, No. 1, pp. 16–30.

Ghadge A, Dani S, and Kalawsky R. *Systems Thinking for Modeling Risk Propagation in Supply Networks*, 2011 IEEE International Conference on Industrial Engineering and Engineering Management, 2011. December 6, pp. 1685–1689, IEEE.

Ghadge, A., Dani, S. and Roy, K. Supply chain risk management: present and future scope, *International Journal of Logistics Management*, 2012. Vol. 23, No. 3, pp. 313–339.

Ghiselin, M.T. Biology, economics, and bioeconomics. In G. Radnitzky (ed.) *Universal Economics: Assessing the Achievements of the Economic Approach*. New York: ICUS Publications, 1992, pp. 55–104.

Ghiselin, M.T. *The Economy of Nature and the Evolution of Sex*. Berkeley: University of California Press, 1974.

Ghobadian, A., Speller, S. and Jones, M. Service quality concepts and models, *International Journal of Quality and Reliability Management*, 1994. Vol. 11, No. 9, pp. 43–66.

Gintis, H., Bowles, S., Boyd, R. and Fehr, E. (eds.). *Moral Sentiments and Material Interests: The Foundations of Cooperation in Economic Life*. Cambridge: MIT Press, 2005.

Givescu, O. The ORESTE' s method in the multicriteria's decision process for the management of tourism field, *Faculty of Management, Academy of Economic Studies, Journal Economia Management*, 2007. Vol. 10, No. 1, pp. 37–51.

Glushkov, V.M., Tseitlin, G.E. and Yushenko, E.L. *Algebra. Language. Programming*. Kiev: Nauka dumka, 1989.

Goh, M., Lim, J.Y.S. and Meng, F. A stochastic model for risk management in global supply chain networks, *European Journal of Operational Research*, 2007. Vol. 182, No. 1, pp. 164–173.

Goldenberg, A. *A Course in Econometrics*. Cambridge, MA: Harvard University Press, 1990.

Gordon, L.A., Loeb, M.P. and Tseng, C.-Y. Enterprise risk management and firm performance: a contingency perspective, *Journal of Accounting and Public Policy*, 2009. Vol. 28, No. 4, pp. 301–327.

Guding D. and Lennoks, Dg., *Weltanschauung: For What Do We Live in the World and What Is Our Place in the World?* Transl. from English, T.V. Barchunoboy (ed.). Yaroslavl': TF Nord, 2001.

Gunasekaran, A., Patel, C. and McGaughey, R.E. A framework for supply chain performance measurement, *International Journal of Production Economics*, 2004. Vol. 87, No. 3, pp. 333–347.

Gupta, A. and Maranas, C.D. Managing demand uncertainty in supply chain planning, *Computers and Chemical Engineering*, 2003. Vol. 27, No. 8–9, pp. 1219–1227.

Hallikas, J., Karvonen, I., Pulkkinen, U., Virolainen, V-M. and Tuominen, M. Risk management processes in supplier networks, *International Journal of Production Economics*, 2004. Vol. 90, No. 1, pp. 47–58.

Hallikas, J., Virolainen, V.M. and Tuominen, M. Risk analysis and assessment in network environments: a dyadic case study, *International Journal of Production Economics*, 2002. Vol. 78, No.1, pp. 45–55.

Hammerstein, P. and Selten, R. Game theory and evolutionary biology. In R.J. Aumann and S. Hart (eds.) *Handbook of Game Theory with Economic Applications*. Amsterdam: Elsevier, 1994, Vol. 2, pp. 929–993.

Han, G. and Shapiro, S. *Statistical Models in Engineering Tasks*. Transl. from English. Moscow: Mir, 1969.

Handfield, R.B. and Ernest L.N. *Supply Chain Redesign: Transforming Supply Chains into Integrated Value Systems*. Upper Saddle River, NJ: FT Press, 2002, pp. 371–372.

Handfield, R.B. and McCormack, K.P. *Supply Chain Risk Management: Minimizing Disruptions in Global Sourcing*. Boca Raton, FL: Taylor and Francis, 2007.

Handfield, R.B. and McCormack, K. *Supply Chain Risk Management: Minimizing Disruptions in Global Sourcing*. New York: Auerbach Publications, Taylor and Francis Group, 2008.

Harland, C., Brenchley, R. and Walker, H. Risk in supply networks, *Journal of Purchasing and Supply Management*, 2003. Vol. 9, No. 2, pp. 51–62.

Hershey, J.C. and Schoemaker, R. Probability versus certainty equivalence methods in utility measurement: Are they equivalent? *Management Science*, 1985. Vol. 31, pp.1213–1231.

Herwijnen, M. van. *Spatial Decision Support for Environmental Management*. Amsterdam: Vrije Universiteit, 1999.

Hinlopen, E., Nijkamp, P. and Rietveld, P. Qualitative discrete multiple-criteria choice models in regional planning, *Regional Science and Urban Economics*, 1983. Vol. 13, pp. 77–102.

Hirshleifer, J. Economics from a biological viewpoint, *Journal of Law and Economics*, April 1977. Vol. 20, No. 1, pp. 1–52.

Hirshleifer, J. Evolutionary models in economics and law: cooperation versus conflict strategies, *Research in Law and Economics*, 1982. Vol. 4, pp. 1–60.

Hobbs, B. F. and Meier, P. *Energy Decisions and the Environment – A Guide to the Use of Multicriteria Methods.* Dordrecht, Netherlands: Kluwer Academic Publishers. 2000.

Hodgson, G.M. *Economics and Evolution: Bringing Back Life into Economics.* Cambridge: Polity Press, 1993.

Hodgson, G.M. Darwinism in economics: from analogy to ontology, *Journal of Evolutionary Economics*, 2002. Vol. 12, No. 3, pp. 259–281.

Hodgson, G.M. *The Evolution of Institutional Economics: Agency, Structure, and Darwinism in American Institutionalism.* London: Routledge, 2004.

Hodgson, G.M. *Biology and Economics: A Very Long Engagement.* A working paper, 2007.

Hsu, C.C., Tan, K.C. and Cross, J. *Influence of resource-based capability and inter-organizational coordination on supply chain management focus*, The 11th International DSI and the 16th APDSI Joint Meeting, Taipei, Taiwan, 2011 July 12–16.

Hu, A.H., Hsu, C.-W., Kuo, T.-C. and Wu, W.-C. Risk evaluation of green components to hazardous substance using FMEA and FAHP, *Expert Systems with Applications*, 2009. Vol. 36, No. 3, pp. 7142–7147.

Hult, G.T.M. and Craighead, C.W. Risk uncertainty and supply chain decisions: a real options perspective, *Decision Sciences*, 2010. Vol. 41, No. 3, pp. 435–458.

Hurley, S. and Nudds, M. (eds.). *Rational Animals?* Oxford: Oxford University Press, 2006.

Hwang, B.-G., Zhao, X. and Yu, G.S. Risk identification and allocation in underground rail construction joint ventures: contractors' perspective, *Journal of Civil Engineering and Management*, 2016. Vol. 22, pp. 758–767.

Hwang, C.L. and Yoon, K. *Multiple Criteria Decision Making. Problems May not Always Have a Conclusive or Unique Solution.* Berlin: Springer-Verlag, 1981.

Ivanchenko, N.N. Researches in the area of work process, gaseous exchange and blast of diesels. In *Dieselestroenie.* Saint Petersburg: Mashinostroenie, 1974.

Ivanishev, V.V. and Marley, V.E. *Introduction in the Theory of Algorithmic Networks.* Saint Petersburg: STU, 2000.

Ivchenko, B.P. and Martishenko, L.A. *Information Ecology.* Saint Petersburg: Nordmed, 1998, 2000, Vol. 1 and 2.

Ivchenko, B.P., Martishenko, L.A. and Monastirskiy, M.L. *Theoretical Bases of Information and Statistical Analysis of Quality of Complex Systems.* Saint Petersburg: Lan', 1997.

Jacquet-Lagrèze, E. and Siskos, Y. Assessing a set of additive utility functions for multicriteria decision making. The UTA method, *European Journal of Operational Research*, 1982. Vol. 10, No. 2, pp. 151–164.

Jammernegg, W. and Reiner, G. Performance improvement of supply chain process by coordinated inventory and capacity management, *International Journal of Production Economics*, 2007. Vol. 108, No. 2, pp. 183–190.

Janssen, R., van Herwijnen, V. and Beinat, E. *DEFINITE for Windows. A system to support decisions on a finite set of alternatives (Software package and user manual).* Amsterdam: Institute for Environmental Studies (IVM), Vrije Universiteit, 2001. (see. https://research.vu.nl/en/publications/definite-for-windows-a-system-to-support-decisions-on-a-finite-se).

Jüttner, U., Peck, H. and Christopher, M. Supply chain risk management: outlining an agenda for future research, *International Journal of Logistics: Research and Applications*, 2003. Vol. 6, No. 4, pp. 197–210.

Kagel, J.H., Battalio, R.C. and Green, L. *Economic Choice Theory: An Experimental Analysis of Animal Behavior*. Cambridge: Cambridge University Press, 1995.

Karlin, S. and Lessard, S. *Theoretical Studies on Sex Ratio Evolution*. Princeton: Princeton University Press, 1986.

Keeney, R. and Raiffa, H. *Decisions with Multiple Objectives. Preferences and Value Tradeoffs*. New York: Wiley, 1976.

Keith, O.R. and Webber, M.D. *Supply-Chain Management: Logistics Catches Up with Strategy*, Outlook. Booz: Allen and Hamilton Inc., 1982.

Keller, E.F. Reproduction and the central project of evolutionary theory, *Biology & Philosophy*, 1987, Vol. 2, pp. 383–396.

Kern, D., Moser, R., Hartmann, E. and Moder, M. Supply risk management: model development and empirical analysis, *International Journal of Physical Distribution & Logistics Management*, 2012. Vol. 42, No. 1, pp. 60–82.

Khalil, E.L. Economics, biology, and naturalism: three problems concerning the question of individuality, *Biology & Philosophy*, April 1997. Vol. 12, No. 2, pp. 185–206.

Khalil, E.L. Survival of the most foolish of fools: the limits of selection theory, *Journal of Bioeconomics*, 2000. Vol. 2, No. 3, pp. 203–220.

Khalil, E.L. *Are Plants Rational? Charles Darwin Meets Organismus Economics*. A working paper, 2007a.

Khalil, E.L. *Why Natural Selection Cannot Explain Rationality*. A working paper, 2007b.

Khan, O., Christopher, M. and Burnes, B. The impact of product design on supply chain risk: a case study, *International Journal of Physical Distribution & Logistics Management*, 2008. Vol. 38, No. 5, pp. 412–432.

Kim, Y.G., Jeong, D., Park, S-H. and Baik, D-K. *Simulation of Risk Propagation Model in Information Systems*, in International Conference on Computational Intelligence and Security, 2006. IEEE Guangzhou, pp. 1555–1558.

Kinney, G.F. and Wiruth, *A.D. Practical Risk Analysis for Safety Management*. China Lake, CA: Naval Weapons Center, 1976 June.

Kleindorfer, P.R. and Saad, G.H. Managing disruption risks in supply chains, *Production and Operations Management*, 2005. Vol. 14, No. 1, pp. 53–68.

Klibi, W. and Martel, A. Scenario-based supply chain network risk modeling, *European Journal of Operational Research*, 2012. Vol. 223, No. 3, pp. 644–658.

Knudsen, T. Economic selection theory, *Journal of Evolutionary Economics*, 2002. Vol. 12, No. 4, pp. 443–470.

Korpela J, Kyläheiko K, Lehmusvaara A, and Tuominen M. An analytic approach to production capacity allocation and supply chain design. *International Journal of Production Economics*, 2002. Vol. 78, No. 2, pp. 187–195.

Koslowski, P. (ed.) *Sociobiology and Bioeconomics: The Theory of Evolution in Biological and Economic Theory*. Berlin: Springer, 1999.

Krasnov, O.V. *Methods and Models of Investigation of Safety of Operation of Rocket Launching Pad*. Saint Petersburg: VISA named A.F. Mojaisky, 2000.

Krasnov, O.V. *Safety Operation of Complex Engineering Systems*. Saint Petersburg: VISA named A.F.Mojaisky, 2002.

Kusimin, I.I., Mahutov, N.A. and Hetagurov, S.V. *Safety and Risk: Ecological and Economic Aspects*. Saint Petersburg: SPbGUEF, 1997.

Labeau, P.E., Smidt, C. and Swaminathan, S. Dynamic reliability: towards an integrated platform for probabilistic risk assessment, *Reliability Engineering and System Safety*, 2000. Vol. 68, pp. 219–254.

Landa, J.T. The political economy of swarming in honeybees: voting-with-the-wings, decision-making costs, and the unanimity rule, *Public Choice*, 1986. Vol. 51, pp. 25–38.

Landa, J.T. and Wallis, A. Socio-economic organization of honeybee colonies: a transaction-cost approach, *Journal of Social and Biological Structures*, July 1988. Vol. 11, No. 3, pp. 353–363.

Lane, D.C. The power of the bond between cause and effect: Jay Wright Forrester and the field of system dynamics, *System Dynamics Review*, 2007. Vol. 23, No. 2–3, pp. 95–118.

Langley, C.J. The evolution of the logistics concept, from logistics. *The Strategic Issues*, Christopher, M. (ed.). 1992.

Lavastre, O., Gunasekaran, A. and Spalanzani, A. Supply chain risk management in French companies, *Decision Support Systems*, 2012. Vol. 52, No. 4, pp. 828–838. Available online 22 November 2011, doi:10.1016/j.dss.2011.11.017.

Levary, R.R. Using the analytic hierarchy process to rank foreign suppliers based on supply risks, *Computers and Industrial Engineering*, 2008. Vol. 55, No. 2, pp. 535–542.

Lindroth, B. and Norrman, A. Categorization of supply chain risk and risk management. In C. Brindley (ed.) *Supply Chain Risk*. Wiltshire: Ashgate Publishing Limited, 2004.

Liu, Z., Lai, M., Zhou, T. and Zhou, Y. *A supply chain risk assessment model based on multistage influence diagram*, In 6th International Conference on Service Systems and Service Management, Xiamen, China, 2009. pp. 72–75.

Lopreato, J. and Crippen, T.A. *The Crisis in Sociology: The Need for Darwin*. New Brunswick: Transaction Publishers, 1999.

Luna-Reyes, L.F. and Andersen, D.L. Collecting and analyzing qualitative data for system dynamics: methods and models, *System Dynamics Review*, 2003. Vol. 19, No. 4, pp. 271–296.

Lysons, K. and Farrington, B. *Purchasing and Supply Chain Management*. Hants: Ashford Colour Press, 2006.

MacArthur, R.H. and Pianka, E.R. On the optimal use of a patchy environment, *American Naturalist*, 1966. Vol. 100, pp. 603–609.

Machalek, R. The evolution of macro society: why are large societies rare? *Advances in Human Ecology*, 1992. Vol. 1, pp. 33–64.

Mahutov, N.A. The problem of risk decreases of beginnings of extraordinary conditions of technogenic nature, *Problems of Safety at Extraordinary Conditions*, 2001. Vol. 3, pp. 29–41.

Mahutov, N.A., Petrov, V.P. and Gadenin, M.M. *Scientific development on integrated problems of Russia safety*. Proceedings of the First International Scientific School: Modelling and Analysis of Safety and Risk in Complex Systems. Saint Petersburg: Omega, 2001, June 18–22.

Manas, M. and Nedoma, J. Finding all vertices of a convex polyhedron, *Numerical Mathematics*, 1968. Vol. 12, pp. 226–229.

Manthou, V., Vlachopoulou, M. and Folinas, D. Virtual e–Chain (VeC) model for supply chain collaboration, *International Journal of Production Economics*, 2004. Vol. 87, No. 3, pp. 241–250.

Manuj, I. and Mentzer, J.T. Global supply chain risk management strategies, *International Journal of Physical Distribution and Logistics Management*, 2008a. Vol. 38, No. 3, pp. 192–223.

Manuj, I. and Mentzer, J.T. Global supply chain risk management, *Journal of Business Logistics*, 2008b. Vol. 29, No. 1, pp. 133–155.

March, J.G. and Shapira, Z. Managerial perspective on risk and risk taking, *Management Science*, 1987. Vol. 33, No. 11, pp. 1404–1418.

Markowitz, H. Portfolio selection. *Journal of Finances*, 1952. Vol. 7, pp. 77–91.

Martel, J.M. and Matarazzo, B. Multiple criteria decision analysis. State of the art surveys, *International Series in Operations Research and Management Science*, 2005. Vol. 78, No. III, pp. 197–259.

Masters, J.M. and Pohlen, T.L. *Evolution of the Logistics Profession. from the Logistics Handbook*, Eds. Roberson, Capcino & Howe, Free Press:New York, 1994.

Masters, R.D. and Gruter, M. (eds.). *The Sense of Justice: Biological Foundations of Law*. Newbury Park: Sage, 1992.

Maynard Smith, J. *The Evolution of Sex*. Cambridge: Cambridge University Press, 1978a.

Maynard Smith, J. Optimization theory in evolution, *Annual Review of Ecology and Systematics*, 1978b. Vol. 9, pp. 31–56.

Mayr, E. *Evolution and the Diversity of Life*. Cambridge, MA: Harvard University Press, 1976.

Mayr, E. *The Growth of Biological Thought: Diversity, Evolution, and Inheritance*. Cambridge, MA: Harvard University Press, 1982.

Mayr, E. *Toward a New Philosophy of Biology*. Cambridge, MA: Harvard University Press, 1988.

McCarthy, D. Liability and risk, *Philosophy and Public Affairs*, 1996. Vol. 25, No. 3, pp. 238–262.

McFarland, D.J. Decision making in animals, *Nature*, 1 September 1977. Vol. 269, pp. 15–21.

McKenna, H.P. The Delphi technique: a worthwhile research approach for nursing? *Journal of Advanced Nursing*, 1994. Vol. 19, No. 6, pp. 1221–1225.

Meixell, M.J. and Gargeya, V. B. Global supply chain: a literature review and critique, *Transportation Research Part E*, 2005. Vol. 41, No. 6, pp. 531–550.

Mele, F.D., Guillen, G., Espuna, A. and Puigjaner, L. An agent-based approach for supply chain retrofitting under uncertainty, *Computer and Chemical Engineering*, 2007. Vol. 31, No. 6, pp. 722–735.

Mentzer, J., DeWitt, W., Keebler, J., Min, S., Nix, N., Smith, C. and Zacharia, Z. Defining supply chain management, *Journal of Business Logistics*, 2001. Vol. 22, No. 2, pp. 1–24.

Mercier, P. and Teghem, J. *Assignment of available products to orders with the MCDM Software Oreste, Applied Mathematics and Computation*, 1993. Vol. 54, No. 2–3, pp. 183–196.

Millstein, R.L. Natural selection as a population-level causal process, *British Journal for the Philosophy of Science*, December 2006. Vol. 57, No. 4, pp. 627–653.

Mojaev, A.S. and Gromov, V.N. *Theoretic Basis of Common Logic and Probabilistic Methods of Automated Modelling Systems*. Saint Petersburg: VITU, 2000.

Mokyr, J. Economics and the biologists: a review of Geerat J. Vermeiy's nature: an economic history, *Journal of Economic Literature*, December 2006. Vol. 44, No. 4, pp. 1005–1013.

Moore, A.F. U2 and the myth of authenticity in rock, *Popular Musicology*, 1998a. Vol. 3, pp. 5–33.

Moore, A.F. In a big country: The portrayal of wide open spaces in the music of big country. In R. Monelle (ed.) *Musica Significans: Proceedings of the 3rd International Congress on Musical Signification*. London: Harwood Academic, pp. 1–6, 1998b.

Mourits, M. and Evers, J.J. Distribution network design. *International Journal of Physical Distribution & Logistics Management*, 1995. Vol. 25, pp. 43–57.

Müller, G.B. and Newman, S.A. (eds.). *Origination of organizational form: beyond the gene in developmental and evolutionary biology*. Cambridge, MA: MIT Press, 2003.

Munda, G. Multi-criteria analysis. In J. Proops and P. Safonov (eds.) *Modelling in Ecological Economics*. Cheltenham: Edward Elgar, NERA (National Economic Research Associates), 2005.

Nagurney, A., Cruz, J., Dong, J. and Zhang, D. Supply chain networks, electronic commerce, and supply side and demand side risk, *European Journal of Operational Research*, 2005. Vol. 164, No. 1, pp. 120–142.

Narasimhan, R. and Talluri, S. Perspectives on risk management in supply chains, *Journal of Operations Management*, 2009. Vol. 27, No. 2, pp. 114–118.

Nelson, R.R. Recent evolutionary theorizing about economic change, *Journal of Economic Literature*, March 1995. Vol. 33, No. 1, pp. 48–90.

Nelson, R.R. and Winter, S.G. *An Evolutionary Theory of Economic Change*. Cambridge, MA: Harvard University Press, 1982.

Norrman, A. and Lindroth, B. *Supply Chain Risk Management: Purchasers' vs. Planners' Views on Sharing Capacity Investment Risks in the Telecom Industry*. Paper presented at the 11th International IPSERA conference, Enschede, The Netherlands. 2002.

Oehmen, J., Ziegenbein, A., Robert, A. and Schönsleben, P. Oriented supply chain risk management, *Production Planning and Control*, 2009. Vol. 20, No. 4, pp. 343–361.

Office of the Deputy Prime Minister (ODPM, Government UK). *DTLR multi-criteria analysis manual. Corporate Publication*. Internet. 2000. www.communities.gov.uk/index.asp?id=1142251

Ohbyung, K., Ghi, P.I. and Lee, K.C. MACE-SCM: a multi-agent and case-based reasoning collaboration mechanism for supply chain management under supply and uncertainties, *Expert Systems with Applications*, 2007. Vol. 33, No. 3, pp. 690–705.

Oke, A. and Gopalakrishnan, M. Managing disruptions in supply chains: a case study of a retail supply chain, *International Journal of Production Economics*, 2009. Vol. 118, No. 1, pp. 168–174.

Olson, DL. *Decision Aids for Selection Problems*. Berlin, Germany: Springer Science & Business Media; 1996.

Ortells, L.E. *Spatial Fuzzy Clustering with Simultaneous Estimation of Markov Random Field Parameters and Class*, PhD thesis, Illinois Institute of Technology Telecom BCN, Polytechnic University of Catalonia. 2011.

Pai, R.R., Kallepalli, V.R., Caudill, R.J. and Zhou, M-C. *Methods toward supply chain risk analysis, in Systems, Man and Cybernetics, 2003*. IEEE International Conference, Washington, DC, 2003. Vol. 5, pp. 4560–4565.

Pak, D., Han, C. and Hong, W.-T. Iterative speedup by utilizing symmetric data in pricing options with two risky assets, *Symmetry*, 2017. Vol. 9, p. 12.

Pastijin, H. and Leysen, J. Constructing an outranking relation with ORESTE, *Mathematical and Computer Modelling*, 1989. Vol. 12, No. 10–11, pp. 1255–1268.

Pechenin, N. K. Concept of managing risk on historical examples, *Alternative energetic and ecology*, 2000. Vol. 1, pp. 116–129.

Peck, H. Drivers of supply chain vulnerability: an integrated framework, *International Journal of Physical Distribution and Logistics Management*, 2005. Vol. 35, No. 4, pp. 210–232.

Peck, H., Abley, J., Christopher, M., Haywood, M., Saw, R., Rutherford, C. and Strathern, M. *Creating Resilient Supply Chains*. Cranfield University, Cranfield School of Management, UK, 2003.

Peniwati, K. "The analytic hierarchy process: The possibility theorem for group decision making." In *Proceedings of the Fourth International Symposium on the Analytic Hierarchy Process*, pp. 202–214. Simon Fraser University Burnaby, 1996.

Pillay, A. and Wang J. *Technology and Safety of Marine Systems*, Bhattacharyya R. and McCormick, M.E. (eds.), Elsevier Ocean Engineering Book Series, 2003. Vol. 7, Elsevier Science Ltd, Oxford, UK.

Pospelov, D.A. *Logic and Linguistic Models in System Control.* WIoscow: Energoizdat, 1976.

PricewaterhouseCoopers. *A practical guide to risk assessment*, 2008. www.pwc.com/en_us/us/issues/enterprise-riskmanagement/ assets/risk_assessment_guide.pdf (accessed 08 April 2013).

Problems of Destruction, Resource and Safety of Engineering Systems. Krasnoyarsk: Siberia, 1997.

Prohorovich, V.E. *Condition forecasting of complex technical systems.* Saint Petersburg: Nauka, 1999.

Raby, C.R., Alexis, D.M., Dickinson, A. and Clayton, N.S. Planning for the future by Western scrub-jays, *Nature*, 22 February 2007. Vol. 445, pp. 919–921.

Rao, S. and Goldsby, T.J. *Supply chain risks: a review and typology, The International Journal of Logistics Management*, 2009. Vol. 20, No. 1, pp. 97–123.

Raphael, D. (ed.). *Social Determinants of Health: Canadian Perspectives.* Toronto, Canada: Canadian Scholars' Press, 2009.

Ravanshadnia, M. and Rajaie, H. Semi-ideal bidding via a fuzzy TOPSIS project evaluation framework in risky environments, *Journal of Civil Engineering Management*, 2013. Vol. 19, No. Suppl. 1, pp. S106–S115.

Rhee, S.J. and Ishii, K. Using cost based FMEA to enhance reliability and serviceability, *Advanced Engineering Informatics*, 2003. Vol. 17, No. 3–4, pp. 179–188.

Ribeiro, C., Ribeiro, A.R., Maia, A.S. and Tiritan, M.E. Occurrence of chiral bioactive compounds in the aquatic environment: a review. *Symmetry*, 2017. Vol. 9, p. 215.

Richardson, G. P. and Pugh III, A.I. *Introduction to System Dynamics Modeling with DYNAMO.* Portland: Productivity Press, 1981.

Risk Management: Risk, Stability Development, Synm-Gistic. (Series Cybernetics: Unlimited Possibilities and Possible Limitations). Moscow: Nauka, 2000.

Roberts, R. and Goodwin, P. Weight approximations in multi-attribute decision models, *Journal of Multicriteria Decision Analysis*, 2002. Vol. 11, pp. 291–303.

Roubens, M. Preference relations an actions and criteria in multicriteria decision making, *European Journal of Operational Research*, 1982. Vol. 10, pp. 51–55.

Roy, B. and Bouyssou, D. An example of a comparison of two decision-aid models. In G. Fandel and J. Spronk (eds.) *Multiple Criteria-Decision Methods and Applications. Selected readings of the First International Summer School Acireale, Sicily, 1985.* Berlin: Springer, September 1983, pp. 361–381.

Russia Safety. *Operation and Development of Complex National Economic Engineering, Energetic, Transport and Communication Systems, v. 1 and 2.* Moscow: Znanie, 1998.

Russia Safety. *Legal, Social and Economic, Research – Engineering Aspects. Dictionary of terms and definitions, v.2.* Moscow: Znanie, 1999.

Ryabinin, I.A. *Reliability of Engineering Systems. Principles and Analysis.* Moscow: Mir, 1976.

Ryabinin, I.A. *Reliability and Safety of Structure-Complex Systems.* Saint Petersburg: Politecknika, 2000.

Ryabinin, I.A. and Cherkesov, G.N. *Logic and Probabilistic Investigation Methods of Structure-Complex Systems Reliability.* Moscow: Radio and communication, 1981.

Saaty, T.L. *The Analytic Hierarchy Process.* New York: McGraw- Hill, 1980.

Sadeghi, N., Fayek, A. and Pedrycz, W. Fuzzy Monte Carlo simulation and risk assessment in construction. *Computer-Aided Civil and Infrastructure Engineering*, 2010. Vol. 25, pp. 238–252.

Sanders, N.R. Pattern of information technology use: the impact of buyer-supplier coordination and performance, *Journal of Operations Management*, 2008. Vol. 26, No. 3, pp. 349–367.

Sankar, N.R. and Prabhu, B.S. Modified approach for prioritization of failures in a system failure mode and effects analysis, *International Journal of Quality & Reliability Management*, 2001. Vol. 18, No. 3, pp. 324–335.

Scarff, F., Carty, A. and Charette, R. *Introduction to the Management of Risk*. Norwich: HMSO, 1993.

Schieg, M. Risk management in construction project management, *Journal of Business Economics Management*, 2006. Vol. 7, pp. 77–83.

Schoener, T.W. Theory of feeding strategies, *Annual Review of Ecology and Systematics*, 1971. Vol. 2, pp. 369–404.

SEC. *Impact Assessment Guidelines*, European Commission, SEC (2005) 791, 15 June 2005.

Shapiro, R.D. Get leverage from logistics, *Harvard Business Review*, 1984. Vol. 62, pp. 119–126.

Sharifi, A. and van Herwijnen, M. *Spatial Decision Support Systems*. International Institute for Geoinformation Science and Earth Observation (ITC) Enschede, 2002.

Sharp, W., Alexander G. and Boily D. *Investigation*. Moscow: Infra, 2001.

Sheffield, J., Sankaran, S and Haslett, T. Systems thinking: taming complexity in project management, *On the Horizon*, 2012. Vol. 20, No. 2, pp. 126–136.

Sheu, C., Lee, L. and Niehoff, B. A voluntary logistics security program and international supply chain partnership, *Supply Chain Management: An International Journal*, 2006. Vol. 11, No. 4, pp. 363–374.

Shokley, J. and Ellis, S.C. *Measuring the Supply Risk Construct: An Exploratory Study*. Clemson: Clemson University, 2006.

Siskos, Y. and Yannacopoulos, D. UTASTAR. An ordinal regression method for building additive value functions, *Investigação Operacional*, 1985. Vol. 5, No. 1, pp. 39–53.

Sislian, E. and Satir, A. Strategic sourcing: a framework and a case study, *The Journal of Supply Chain Management*, 2000. Vol. 36, No. 3, pp. 4–11.

Skulmoski, G.J., Hartman, F.T. and Krahn, J. The Delphi method for graduate research, *Journal of Information Technology Education*, 2007. Vol. 6, No. 0, pp. 1–21.

Skurihin, V.I. and Morosov, A.A. Integrated automation systems of management. Features of construction and directions of development. *Controlling systems and machines*, 1976. Vol. 2, pp. 5–11.

Smeltzer, L.R. and Siferd, S.P. Proactive supply management: The management of risk, *The Journal of Supply Chain Management*, 1998. Vol. 34, No. 1, pp. 38–45.

Smith, E.A. and Winterhalder, B. Natural selection and decision-making: some fundamental principles. In E.A. Smith and B. Winterhalder (eds.) *Evolutionary Ecology and Human Behavior*. New York: Aldine de Gruyter, 1992, pp. 25–60.

Sober, E. Three differences between deliberation and evolution. In P.A. Danielson (ed.) *Modeling Rationality, Morality, and Evolution*. New York: Oxford University Press, 1998, pp. 408–422.

Sodhi, M.S. and Lee, S. An analysis of sources of risk in the consumer electronics industry, *Journal of Operation Research Society*, 2007. Vol. 58, No. 11, pp. 1430–1439.

Sodhi, M.S., Son, B.G. and Tang, C.S. Perspectives on supply chain risk management, *International Journal of Production and Operations Management*, 2012. Vol. 21, No. 1, pp. 1–13.

Sodhi, M.S. and Tang, C.S. *Strategic Approaches for Mitigating Supply Chain Risks*, Managing Supply Chain Risk. New York: Springer, 2012, pp. 95–107.

Sohn, S.Y. and Lim, M. The effect of forecasting and information sharing in SCM for multi-generation products, *European Journal of Operational Research*, 2008. Vol. 186, No. 1, pp. 276–287.

Solnitsev, R.I. *Computer-Aided Design of Systems of Automation Control.* Moscow: Height School, 1991.

Solojentsev, E.D. *Bases of System Development for Automated Debugging of Complex Objects of Machine-Buildings*, The thesis of Dr. Sc. Kiev, Institute of Cybernetic. 1982.

Solojentsev, E.D. *Introduce in Intellectual AWS and Expert Systems in Machine-Building; Textbook.* Saint Petersburg: LIAP, 1991.

Solojentsev, E.D. Method of automated debugging of complex objects volumetric energetic machines, *Automation*, 1981. Vol. 2, pp. 68–74.

Solojentsev, E.D. and Alekseev, V. *Logic and probabilistic theory of security portfolio risk.* Proceedings of the Third International Scientific School: Modelling and Analysis of Safety and Risk in Complex Systems; 2003 August 20–23; Saint Petersburg: SPbSUASI, 2003.

Solojentsev, E.D. and Karassev, V.V. Logic and probabilistic risk models in business with groups of incompatible events. *Economics and Mathematics Methods*, 2003. Vol. 1, pp. 90–105.

Solojentsev, E.D., Karassev, V.V. and Solojentsev, V.E. *Logic and Probabilistic Models of Risk in Banks, Business and Quality.* Saint Petersburg: Nauka, 1999.

Spekman, R.E. and Davis, D.E. Risky business: expanding the discussion on risk and the extended enterprise, *International Journal of Physical Distribution and Logistics Management*, 2004. Vol. 34, No. 5, pp. 414–433.

Stank, T.P., Keller, S.B. and Daugherty, P.J. Supply chain collaboration and logistical service performance, *Journal of Business Logistics*, 2001. Vol. 22, No. 1, pp. 29–48.

Stecke, K.E. and Kumar, S. Sources system of supply chain disruptions, factors that breed vulnerability, and mitigating strategies, *Journal of Marketing Channels*, 2009. Vol. 16, No. 3, pp. 193–226.

Stephen, C.H.L., Sally, O.S.T., Ng, W.L. and Yue, W. A robust optimization model for multi-site production planning problem in an uncertain environment, *European Journal of Operational Research*, 2007. Vol. 181, No. 1, pp. 224–238.

Stephens, D.W. and Krebs, J.R. *Foraging Theory.* Princeton, NJ: Princeton University Press, 1986.

Sterman, J.D. *Business Dynamics: Systems Thinking and Modelling for a Complex World.* Boston, MA: Irwin/McGraw-Hill, 2000.

Stevens, G.C. Integrating the supply chain, *International Journal of Physical Distribution & Materials Management*, 1989. Vol. 19, pp. 3–8.

Sato, S. and Kumamoto, H. *Reengineering the Environment.* New York: Vantage Press, 1995.

Swaminathan, J.M., Smith, S.F. and Zadeh, N.M. Modelling supply chain dynamics: a multi-agent approach, *Decision Sciences*, 1998. Vol. 29, No. 3, pp. 607–632.

Szwejczewski, M., Mitchell, R. and Lemke, F. Risk measurement and management during new product development: an exploratory study, *International Journal of Risk Assessment and Management*, 2008. Vol. 9, No. 3, pp. 277–287.

Tang, C. Perspectives in supply chain risk management, *International Journal of Production Economics*, 2006. Vol. 103, No. 2, pp. 451–488.

Tang, C.S. and Tomlin, B. The power of flexibility for mitigating supply chain risks, *International Journal of Production Economics*, 2008. Vol. 116, No. 1, pp. 12–27.

Tang, O. and Nurmaya Musa, S. Identifying risk issues and research advancements in supply chain risk management, *International Journal of Production Economics*, 2010. Vol. 133, No. 1, pp. 25–34.

Taylan, O., Bafail, A., Abdulaal, R. and Kabli, M. Construction projects selection and risk assessment by fuzzy AHP and fuzzy TOPSIS methodologies, *Applied Soft Computing*, 2014, Vol. 17, pp. 105–116.

Thomas, D.J. and Griffin, P.M. Coordinated supply chain management, *European Journal of Operational Research*, 1996. Vol. 94, pp. 1–15.

Thomas, K. and David, C. The risk of second tier suppliers in serial supply chains: implications for order policies and distributor autonomy, *European Journal of Operational Research*, 2008. Vol. 186, No. 3, pp. 1158–1174.

Towill, D.R. The impact of business policy on bullwhip induced risk in supply chain management, *International Journal of Physical Distribution & Logistics Management*, 2005. Vol. 35, No. 8, pp. 555–575.

Trkman, P. and McCormack, K. Supply chain risk in turbulent environments—a conceptual model for managing supply chain network risk, *International Journal of Production Economics*, 2009. Vol. 119, No. 2, pp. 247–258.

Tullock, G. The coal tit as a careful shopper, *American Naturalist*, 1971. Vol. 105, pp. 77–80.

Tullock, G. *The Economics of Non-Human Societies*. Tucson: Pallas Press, 1994.

Tummala, R and Schoenherr, T. Assessing and managing risks using the supply chain risk management process (SCRMP), *Supply Chain Management: An International Journal*, 2011. Vol. 16, No. 6, pp. 474–483.

Tuncel, G. and Alpan, G. Risk assessment and management for networks: a case study, *Computer in Industry*, 2010. Vol. 61, No. 3, pp. 250–259.

Tupolev, A.N. *Verge of Venture Creation*. Moscow: Nauka, 1988.

Vafadarnikjoo, A., Mobin, M. and Firouzabadi, S. *An intuitionistic fuzzy-based DEMATEL to rank risks of construction projects*. In Proceedings of the 2016 International Conference on Industrial Engineering and Operations Management, Detroit, MI, USA, 23–25 September 2016; pp. 1366–1377.

Van de Panne, C. *Methods for Linear and Quadratic Programming*. Amsterdam: North-Holland Publishing Company, 1975.

Van Leeuwen, J.F., Nauta, M.J., de Kaste, D., Odekerken–Rombouts, Y.M.C.F., Oldenhof, M.T., Vredenbregt, M.J. and Barends, D.M. Risk analysis by FMEA as an element of analytical validation, *Journal of Pharmaceutical and Biomedical Analysis*, 2009. Vol. 50, No. 5, pp. 1085–1087.

Vasiliev, V.D. and Solojentsev, E.D. *Cybernetic Methods at Development of Piston Machines*. Moscow: Mashinostroenie, 1978.

Vennix, J.A. and Vennix, J. *Group Model Building: Facilitating Team Learning Using System Dynamics*. Chichester: John Wiley & Sons, 1996.

Vermeij, G.J. *Nature: An Economic History*. Princeton, NJ: Princeton University Press, 2004.

Vincent, T.L. and Brown, J.S. The evolution of ESS theory. *Annual Review of Ecology and Systematics*, 1988. Vol. 19, pp. 423–443.

Vincke, P. Multi-attribute utility theory as a basic approach. In G. Fandel and J. Spronk (eds.) *Multiple Criteria-Decision Methods and Applications. Selected readings of the First International Summer School Acireale, Sicily, 1985*. Berlin: Springer, September 1983, pp. 27–40.

Vladimirov, V.A, Vorob'ev, Y.L. and Salov, S.S. and etc. *Risk Management*. Moscow: Nauka, 2000.

Von Neumann, J. and Morgenstern, O. *Theory of Games and Economic Behaviour*, Princeton, NJ: Princeton University Press, 1947.

Von Winterfeldt, D. and Edwards, W. *Decision Analysis and Behavioural Research*. Cambridge: Cambridge University Press, 1986.

Voogd, H. *Multicriteria Evaluation for Urban and Regional Planning*. London: Pion, 1983.

Vreeker, R., Nijkamp, P. and Ter Welle, C. A multicriteria decision support methodology for evaluating airport expansion plans, *Transportation Research Part D*, 2002. Vol. 7, pp. 27–47.

Wagner, S.M. and Neshat, N. Assessing the vulnerability of supply chains using graph theory, *International Journal of Production Economics*, 2010. Vol. 126, No. 1, pp. 121–129.

Wang, X., Chan, H., Yee, R. and Diaz-Rainey, I. A two-stage fuzzy-AHP model for risk assessment of implementing green initiatives in the fashion supply chain, *International Journal of Production Economics*, 2012. Vol. 135, No. 2, pp. 595–606.

Wang, Y.M., Chin, K.S., Poon, G.K. and Yang, J.B. Risk evaluation in failure mode and effects analysis using fuzzy weighted geometric mean, *Journal of Expert Systems with Applications*, 2009. Vol. 36, No. 2, pp. 1195–1207.

Waters, D. *Supply Chain Risk Management: Vulnerability and Resilience in Logistics*. London: Kogan Page Limited, 2007.

Weibull, J.W. *Evolutionary Game Theory*. Cambridge, MA: MIT Press, 1995.

Winterhalder, B. Analyzing adaptive strategies: human behavioral ecology at twenty-five, *Evolutionary Anthropology*, 2000. Vol. 9, pp. 51–72.

Winterhalder, B. Evolutionary ecology and the social sciences. In E.A. Smith and B. Winterhalder (eds.) *Evolutionary Ecology and Human Behavior*. New York: Aldine de Gruyter, 1992, pp. 3–23.

Winterhalder, B. and Smith, E.A. (eds.). *Hunter-Gatherer Foraging Strategies: Ethnographic and Archeological Analyses*. Chicago, IL: University of Chicago Press, 1981.

Witt, U. *The Evolving Economy: Essays on the Evolutionary Approach to Economics*. Cheltenham: Edward Elgar, 2003.

Wold, G.H. and Shriver, R.F. Risk analysis techniques, *Disaster Recovery Journal*, 1997. Vol. 7, No. 3, pp. 1–8.

Wu, D. and Olson, D. Supply chain risk, simulation, and vendor selection, *International Journal of Production Economics*, 2008. Vol. 114, No. 2, pp. 646–655.

Wu, T., Blackhurst, J. and Chidambaram, V. A model for inbound supply risk analysis, *Computers in Industry*, 2006. Vol. 57, No. 4, pp. 350–365.

Xia, D. and Chen, B. A comprehensive decision–making model for risk management of supply chain, *Expert Systems with Applications*, 2011. Vol. 38, No. 5, pp. 4957–4966.

Xiao, J. J., Tang, C., Serido, J., and Shim, S. Antecedents and consequences of risky credit behavior among college students: Application and extension of the Theory of Planned Behavior. *Journal of Public Policy & Marketing*, 2011. Vol. 30, No. 2, pp. 239–245.

Xiao, T., Xiangtong, Q. and Gang, Y. Coordination of supply chain after demand disruptions when retailers compete, *International Journal of Production Economics*, 2007. Vol. 109, No. 2, pp. 162–179.

Xiao, T. and Yang, D. Price and service competition of supply chains with risk-averse retailers under demand uncertainty, *International Journal of Production Economics*, 2008. Vol. 114, No. 1, pp. 187–200.

Xu, S. *References on the Analytic Hierarchy Process*. Tianjin: Institute of Systems Engineering, Tianjin University, 1986.

Yang, J., Qi, X. and Yu, G. Disruption management in production planning, *Naval Research Logistics*, 2005. Vol. 52, No. 5, pp. 420–442.

Yusupov, R.M. and Zabolotsky, V.P. *Scientific and nzethodical Bases of Informatics*. Saint Petersburg: Nauka, 2002.

Zhelesnuakov, A.B. *Soviet Cosmonautics: Chronicle of Events and Catastrophes*. Saint Petersburg: Nauka, 1998.

Zsidisin, G.A. Managerial perceptions of supply risk, *Journal of Supply Chain Management*, 2003. Vol. 39, No. 1, pp. 14–25.

Index

Note: **Bold** page numbers refer to tables, *italic* page numbers refer to figures and page numbers followed by "n" denote endnotes.